MEDIA PERSPECTIVES ON
INTELLIGENT DESIGN AND EVOLUTION

MEDIA PERSPECTIVES ON
INTELLIGENT DESIGN AND EVOLUTION

Mark Paxton

 GREENWOOD

AN IMPRINT OF ABC-CLIO, LLC
Santa Barbara, California • Denver, Colorado • Oxford, England

Library of Congress Cataloging-in-Publication Data

Paxton, Mark.
  Media perspectives on intelligent design and evolution / Mark Paxton.
    p. cm.
  Includes bibliographical references and index.
    ISBN 978–0–313–38064–8 (cloth : alk. paper) — ISBN 978–0–313–38065–5 (ebook) 1. Intelligent design (Teleology). 2. Creationism. 3. Human evolution—Religious aspects—Christianity. 4. Mass media—Religious aspects—Christianity. I. Title.
BL263.P39   2013
231.7'652—dc23   2012031755

ISBN: 978–0–313–38064–8
EISBN: 978–0–313–38065–5

17 16 15 14 13   1 2 3 4 5

This book is also available on the World Wide Web as an eBook.
Visit www.abc-clio.com for details.

Greenwood
An Imprint of ABC-CLIO, LLC

ABC-CLIO, LLC
130 Cremona Drive, P.O. Box 1911
Santa Barbara, California 93116-1911

This book is printed on acid-free paper ∞

Manufactured in the United States of America

Contents

Introduction

When Charles Darwin took his voyage on the HMS *Beagle* from 1831 to 1836 and visited the Galapagos Islands off the coast of Chile, no one could have envisioned what would transpire after his return. It is not news that Darwin observed wildlife on each of the islands with physical attributes that, he said, could not be explained except through evolution. It also is not news that Darwin's theory of evolution, which he developed based on his visit to Galapagos, has prompted opposition dating all the way back to the publication of his *Origin of Species* in 1859. Darwin's theory reshaped the way people look at the natural world, and it has never ceased to generate controversy.

Much of the opposition to the theory of evolution has been from religion leaders, who contend that the theory that species, including humans, evolved from lower forms is a concept that devalues human life and disregards the Biblical depiction of God creating mankind. But new criticism has emerged since about 1990, when the intelligent design movement was jump-started by publication of two books. One was a science textbook, *Of Pandas and People*, first published in 1989, which was based on the idea that evolution alone cannot explain many aspects of life. Second was the publication of law professor Phillip Johnson's criticism of Darwin's theory in the 1991 book *Darwin on Trial*.

Supporters of this intelligent design movement generally have failed to gain acceptance in the public schools, the result of several high profile court cases that have found intelligent design to be little more than an attempt to insert religious beliefs into education in violation of the First Amendment's prohibition of the establishment of religion. But these intelligent design advocates, supported by the Seattle-based conservative think tank The Discovery Institute, have waged a high-profile media campaign to win public support of their efforts.

The Discovery Institute and its intelligent design arm, the Center for Science and Culture, have taken several approaches to winning public approval. In one approach, intelligent design supporters argue that there is controversy over the theory of evolution and public school students should be taught about this controversy and allowed to make up their own minds about how life developed. A second approach is that intelligent design is a legitimate scientific approach and should have equal footing with other scientific approaches, including the theory of evolution. A third approach is that teachers have the academic freedom to teach subject matter in the way they think best conveys that information to students, and if science teachers want to teach that the theory of evolution is problematic and intelligent design is a legitimate alternative, then they should be free to teach that in class. Above all, intelligent design proponents contend that their approach differs significantly from what is known as creationism—the idea that God alone is responsible for life on Earth as described in the Bible. Instead, intelligent design advocates describe their position as agnostic—they say they do know who the designer was, merely that there is ample evidence that a designer was responsible for some life forms.

Most mainstream scientists, however, argue that intelligent design is just a new version of creationism and that intelligent design proponents are simply trying to insert religion into the public schools. They contend that evolution is the only accepted scientific approach to the development of life, and the only controversy that exists over evolution was created by intelligent design proponents looking for a way to win support for their approach.

The news media have been key players in the dispute between evolution supporters and intelligent design proponents. The classroom or textbook description is that journalists are neutral and objective, and merely relay reality to the reading or viewing audience. In this version of journalism, reporters are simply chroniclers of current events. One journalism textbook, for instance (Harrower, 2007), says that reporters must respect the "integrity of facts" and must carefully report information without inserting their own opinions (34). Another journalism textbook (Ryan ands Tankard, 2005) notes that reporters must use an objective approach, transcending their own personal biases, and that by doing so, they "can describe reality with reasonable accuracy" (20).

But a new approach to analyzing news coverage has developed since the 1980s. Media researchers have begun to examine how the media frame, or depict, events. Rather than being neutral observers of facts, journalists are choosing—whether consciously or unconsciously—what aspect of a news event they will focus on. For example, is a tax on the value of a dead person's assets an "estate tax," as framed by supporters, or a "death tax," as framed by opponents? How that issue is framed helps to win public support, either for the tax or against it. Framing allows the public to make sense of a complex issue such as the tax by providing a kind of road map. The key helps to make a map understandable—the symbols and numbers on a map have no intrinsic value of their own. Similarly, a complex issue needs a "key," or commonly understood ideas, to help readers to make sense of the news.

Studies by communication and political science researchers who examine media framing of issues have found that media consumers depend on frames to help them make sense of complex issues. Intelligent design and evolution is just such a complex issue. Intelligent design proponents cite evidence such as the bacterium flagellum, which they view as irreducibly complex—meaning that all parts are necessary for the flagellum to work, and evolution of the individual parts would have rendered the flagellum inoperable—to argue that a designer must have created the device. Evolution supporters contend that the flagellum did in fact evolve, and the parts that intelligent design supporters cite as irreducible actually are not irreducible. It is safe to say that a vast majority of the public not only is unaware of the bacterium flagellum, but also lacks the scientific background to understand what the debate is about. (The flagellum debate is more fully explained in Chapter 1.)

Framing explains why and how the complex issues of intelligent design and evolution debate are covered by print, broadcast, and Internet reporters. The frames used by journalists when covering the intelligent design-evolution debate fall into several broad categories. One is the framing of the attempt to have schools teach intelligent design as a new version of the 1925 Scopes "monkey" trial, in which a Dayton, Tennessee, school teacher was tried and convicted of violating Tennessee state law prohibiting teachers from discussing evolution in the classroom, a courtroom battle popularized on stage and in the movies by *Inherit the Wind*. (The conviction was later overturned on appeal on the technical issue that the jury and not the judge should have levied the $100 fine against Scopes.) A second common frame that the media use is that intelligent design efforts are really creationism in disguise; a third common media frame is to cast the debate in political terms, as a fight between cultural conservatives and liberals.

This book examines several aspects of the intelligent design-evolution debate and the media coverage of that debate. Chapter 1 takes a look at the history and development of the intelligent design movement, including some of the major arguments supporters make on its behalf, as well as some of the more basic criticisms of the intelligent design movement.

Chapter 2 looks at the argument on behalf of the theory of evolution, tracing the idea's development from the 1800s, through Darwin's two major works—*Origin of Species* and *The Descent of Man*—and how the theory of evolution has developed over the past 150-plus years. This chapter also discusses some of the criticisms leveled at evolution by intelligent design advocates.

The third chapter examines the various efforts to persuade local and state boards of education to adopt policies questioning the theory of evolution as unproven and proposing intelligent design as an alternative approach to teaching about the origin of life. This chapter includes discussion of a number of state and federal court cases on the issue of evolution and intelligent design, including *Kitzmiller v. Dover*, the well-publicized case in Dover, Pennsylvania, that was decided in 2005.

Chapter 4 considers how the media have framed intelligent design and evolution in print, over the air, and on the Internet. It begins with an in-depth discussion of what framing is and how media framing works. It then takes a look at what researchers have found when they have studied framing and the intelligent design-evolution discussion, followed by the author's own study of newspaper reporting and opinion writing about the issue. It then examines television and Internet coverage of the issue.

Chapter 5 provides a conclusion, offering informed speculation about where the debate and media coverage of that debate will go in the coming years. Following the final chapter are several items that can help those interested in further information about intelligent design and evolution, and media coverage of the two ideas. First is a section of brief biographies of the players in the intelligent design-evolution controversy. Next are both a traditional bibliography and an annotated bibliography providing an expanded description of the major works on the subjects of intelligent design, evolution, media framing, and media coverage.

This book is not focused solely on either intelligent design or the theory of evolution. There are plenty of books that focus on one or the other, as evidenced by the bibliographies at the end of this manuscript. Instead, this book provides a relatively brief overview of intelligent design and evolution and the criticisms of both so that readers can grasp the context of the debate, as well as how the news media cover the issue.

As you read this book, you should keep the following things in mind. Whether you are an intelligent design supporter, an evolution supporter, or have not made up your mind yet, pause to consider where you have received most of your information about these issues. It is a good bet that your knowledge came from news media accounts. Next, consider what these media reports focused on, or framed. Then examine whether those frames helped to shape your opinions about intelligent design and the theory of evolution.

CHAPTER 1

Intelligent Design

Since the 1980s, the idea of intelligent design has been at the forefront of the dispute over whether evolution is a valid explanation for the origin of life. Groups such as the Discovery Institute in Seattle, founded in 1990, began to argue that life is too complex not to have been designed by an intelligent creator. According to the Discovery Institute, intelligent design is an approach that combines science and philosophy to seek evidence of design in nature; the Institute argues that "[t]he theory of intelligent design holds that certain features of the universe and of living things are best explained by an intelligent cause, not an undirected process such as natural selection" ("Definition of Intelligent Design," 2010).

Critics contend that intelligent design is merely a dressed up version of creationism, the belief that God created life as described in the Book of Genesis in the Bible. Many evolution supporters ridicule intelligent design as "creation lite." Advocates of intelligent design, however, contend that their approach is just as scientific as the theory of evolution and argue that it provides answers to biological organisms that the theory of evolution cannot answer. This chapter will take an in-depth look at the development of the intelligent design movement, as well as criticisms of intelligent design by most mainstream scientists.

INTELLIGENT DESIGN v. CREATIONISM

In *Debating Design* (2004), William Dembski of the Discovery Institute and his co-author Michael Ruse write that intelligent design proponents believe that evolution alone, or even evolution started by extra-terrestrial beings, cannot explain life. Indeed, Dembski and Ruse state that the concept of intelligent design is based on the idea "that there must be something more than ordinary natural causes or material mechanisms" responsible for life on earth, "and moreover, that something must be intelligent and capable of bringing about organisms" (3). Some intelligent design proponents even suggest that belief in intelligent design does not require disbelief in evolution. Intelligent design supporters do not necessarily believe that "a personal God" is the designer in question because intelligent design has a broader playing field than creationism, which is bound by the concept of God. David K. DeWolf, writing in *The University of St. Thomas Journal of Law Public Policy* (2009), argues that intelligent design does not rely on belief in God but is "agnostic" in that proponents take no position on who the designer was (347).

Other intelligent design proponents disagree, however. For instance, former journalist and self-proclaimed former atheist-turned Christian writer Lee Strobel, in his pro-intelligent design book *The Case for a Creator* (2004), recounts his personal search for scientific evidence for the existence of God and determines that the intelligent design argument depends on the Christian God as the designer. "Unlike Darwinism, where my faith would have to swim upstream against the strong current of evidence flowing the other way," he writes, "putting my trust in the God of the Bible was nothing less than the most rational and natural decision I could make" (285).

Dembski, the unofficial leader of the intelligent design movement, has authored a number of books, book chapters, web postings, and other writings that explain and advocate intelligent design. Dembski earned a bachelor's degree in psychology, a master's degree in statistics, a Ph.D. in philosophy at the University of Illinois at Chicago, and a Ph.D. in mathematics from the University of Chicago. He taught at Northwestern University, the University of Notre Dame, the University of Dallas, Baylor University, and as of 2010, was a professor at Southwestern Baptist Theological Seminary in Fort Worth, Texas.

Dembski frames intelligent design as a scientific approach that studies the signs of intelligence in biological systems, though he says it is not an attempt to determine who the designer was or what that designer was thinking. As such, it is a direct challenge to the Darwinian concept of evolutionary development. Intelligent design supporters cite the complexity of life forms as evidence of a designer (2004, 33–34).

CREATIONISM

An exploration of what intelligent design is must begin with what its supporters say it is not—that is, creationism, or the belief that God created life as it exists today. Creationists fall into two basic camps. One group contends that the description of God creating the heavens and the Earth contained in the Biblical account in Genesis is literally true. Supporters of this approach contend that the Earth is not ancient, as most mainstream scientists contend, but rather that it is only 6,000 to 10,000 years old.

An example of this approach is apparent in the Creation Museum, a 70,000-square-foot facility that opened in Petersburg, Kentucky, a suburb of Cincinnati, Ohio, in 2007. The museum, built at a cost of $27 million, claims to "[bring] the pages of the Bible to life, casting its characters and animals in dynamic form and placing them in familiar settings. Adam and Eve live in the Garden of Eden. Children play and dinosaurs roam near Eden's Rivers" ("About the Museum," 2010). A review of the museum, published in *The New York Times*, describes it in unflattering terms:

> Outside the museum scientists may assert that the universe is billions of years old, that fossils are the remains of animals living hundreds of millions of years ago, and that life's diversity is the result of evolution by natural selection. But inside the museum the Earth is barely 6,000 years old, dinosaurs were created on the sixth day, and Jesus is the savior who will one day repair the trauma of man's fall. It is a measure of the museum's daring that dinosaurs and fossils—once considered major challenges to belief in the Bible's creation story—are here so central, appearing not as tests of faith, as one religious authority once surmised, but as creatures no different from the giraffes and cats that still walk the earth. Fossils, the museum teaches, are no older than Noah's flood; in fact dinosaurs were on the ark. (Rothstein, 2004, E1)

Other creationists believe the Genesis description of God creating the Earth in six days is a metaphor, that no one today knows how long a day was for God—a day might equal epochs, and nothing in the Biblical account means that the Earth is literally young. One of those who believe in old-Earth creationism is Hugh Ross, who through his website www.reasons.org as well as through his books espouses the belief that God created mankind, but that the age of the Earth is millions, if not billions, of years old, as paleontologists and other scientists maintain. He writes that Genesis says that neither islands nor continents existed on early Earth, and that God created most of the land mass on Day 3 of creation. As Ross, an astronomy Ph.D. and writer about Christianity, contends (2008), "The problem with these...passages was that in the early 1960s the prevailing view among geologists and geophysicists was that continental landmasses, though shifting in position and shape, had covered a large fraction of Earth's surface throughout Earth's history. This view, though troublesome, did not disturb my faith in an inerrant Bible."

Intelligent design proponents say their approach follows neither of the creationist scenarios: that God created the Earth in six days (and rested on the seventh), or that God's creation occurred over hundreds of millions of years, if not billions of years. Instead, proponents such as Dembski say they do not know who the designer was. Instead, they infer the designer's existence by examining the biological record for signs of a designer. You can have intelligent design without creationism, intelligent design supporters argue. Dembski writes that creationists are concerned with who started life, while intelligent design supporters are concerned with finding patterns in biologic samples that point to a designer.

Critics, however, cite instances in which Dembski himself, despite stating that no one knows who the designer was, has said that some of intelligent design is attributable to the Christian God. In a 2004 speech to the Fellowship Baptist Church of Waco, Texas, Dembski stated, "[A]nother thing we need to be aware of is that not every instance of design we see in nature needs to be directly attributed to God" (quoted in Coyne, 2006, 17). In fact, however, Dembski's book *The Design Revolution* states unequivocally that Dembski himself is a Christian, and he notes, "As a Christian I hold that the Christian God is the ultimate source of design behind the universe" (26).

More telling, perhaps, is the existence of the "Wedge Document," a Discovery Institute internal document that was leaked on the Internet in 1999. This document outlines the way in which the intelligent design movement plans to replace materialism—the idea that all that matters is the natural world—with a Christian approach. For intelligent design proponents, one of the flaws of the theory of evolution is its focus on the natural world–the material world—as opposed to other spiritual world.

This Wedge Document, which the Discovery Institute acknowledges and defends, lists a five-year strategy. The first phase consists of scientific writing, research, and publicity about intelligent design. The second phase involves publicity and attempts to sway public opinion "to reverse the stifling dominance of materialist worldview, and to replace it with a science consonant with Christian and theistic convictions." The third phase includes seeking legal assistance in response to efforts to integrate intelligent design in public schools. The document states that the intelligent design movement's "natural constituency" is Christian (Foster, Clark, and York, 2008, 38–40). In a lengthy response posted on the Discovery Institute's webpage, the Institute acknowledged that many of the Institute's staff fellows believe in God, but called criticism over the Wedge Document "paranoia" and categorically denied trying to create a Christian approach to science ("The 'Wedge Document': So What," 2006).

HISTORY OF INTELLIGENT DESIGN

Despite negative publicity surrounding the Wedge Document and other criticisms, the Discovery Institute has been a leader in the intelligent design movement. Above all, Dembski and other proponents say, intelligent design is a scientific approach to the origin of life because it is based on empirical, or observable, examples of design in nature. The designer, he writes, is an "intelligent agency."

"In determining whether biological organisms exhibit specified complexity," Dembski writes, "design theorists focus on identifiable systems—such as individual enzymes, metabolic pathways, molecular machines, and the like" (2004, 35). Dembski uses the metaphor of archery, arguing that evolution supporters wait until the archer fires his arrow and then draw a bull's-eye around it, imposing "a pattern after the fact." Intelligent design proponents, he suggests, draw the bull's-eye first, and when the arrow hits the target, say it is evidence of design (35). Dembski also uses the metaphor of Mount Rushmore, the South Dakota monument bearing the carved likenesses of Presidents Washington, Jefferson, Lincoln, and Theodore Roosevelt. Mount Rushmore, he writes, shows evidence of a designer in the same way that naturally observable biological phenomena show signs of a creator (33).

In his web posting "A Brief History of Intelligent Design" (2006), Casey Luskin traces the debate over intelligent design as far back as Socrates and Plato and other Greek philosophers such as Heraclitus and Democritus. Luskin writes that Heraclitus and Democritus believed that life could begin without any intelligent guidance, while Socrates and Plato believed a mind was required to explain life. Sir

Isaac Newton questioned whether eyes and ears could have been developed without some intelligent being's intervention, and critics of Darwin's time claimed, in works of nineteenth-century paleontologist and creationist Louis Agassiz, an intellectual power to explain the wide range of life forms.

One of the sources often cited by intelligent design proponents was written by William Paley, who lived from 1743 to 1805. In his book *Existence and Attributes of the Deity, collected from Appearances of Nature* (1802), Paley imagines walking along a heath and stumbling across a rock; he has no reason to believe the rock had not been there forever. But then, he wrote, what if he stumbled across a watch on the ground? He would have to believe the watch did not occur naturally, that a watchmaker had designed it. Paley's watchmaker served as a theoretical underpinning to the intelligent design movement that would arise almost 200 years later.

In 1982, Fred Hoyle and N. C. Wickremasinghe wrote that biologic evidence had to be the result of intelligent design (1982). Two years later, chemist Charles Thaxton, engineering professor Walter Bradley, and chemist Roger Olsen began the modern intelligent design movement with their book *The Mystery of Life's Origin*, which argued for an "intelligent cause" in the origin of the information in DNA. Thaxton would later be one of the creators of the *Of Pandas and People* text, first published in 1989, which uses intelligent design to question the veracity of evolution. In 1985, molecular biologist Michael Denton published *Evolution: A Theory in Crisis*, which said intelligent design is a legitimate explanation for many things found in nature.

Most observers, however, trace the popularization of the intelligent design movement not to the 1980s but to the 1990s and specifically to Berkeley law professor Philip Johnson and his book *Darwin on Trial* (1991). Johnson was a lifelong agnostic who, after the breakup of his marriage and subsequent remarriage to a devout Christian, began to reconsider Christianity. Along with his conversion, Johnson began to reconsider the idea of evolution. Although he had no training in the sciences, Johnson decided to write as if we were preparing a legal brief against evolution (Hume, 2007, 64–67). In *Darwin on Trial*, Johnson put Darwinism and evolution "on trial" and found them guilty of failing to explain adequately the complexity and adaptation of life. Johnson's argument was that scientists, rather than pursuing all possible explanations for anomalies in the evolution timeline, instead ruled out the possibility of supernatural influence on the development of life. "The question," he wrote, is not whether natural selection occurs. Of course it does, and it has an effect in maintaining the genetic fitness of a population. Infants with severe birth defects do not survive to maturity without expensive medical care, and creatures which do not survive to reproduce do not leave descendants. These effects are unquestioned, but Darwinism asserts a great deal more than merely that species avoid genetic deterioration due to natural attrition among the genetically unfit. Darwinists claim that this same force of attrition has a building effect so powerful that it can begin with a bacterial cell and gradually craft its descendants over billions of years to produce such wonders as trees, flowers, ants, birds, and humans. How do we know that all this is possible? (11)

For Johnson, a perfect example of the impossibility of natural selection to create new specifics is evident in the breeding of animals. "Natural selection is a conservative force that prevents the appearance of the extremes of variation that human breeders like to encourage," he writes.

What artificial selection actually shows is that there are definite limits to the amount of variation that even the most highly skilled breeders can achieve. Breeding of domestic animals has produced no new species, in the commonly accepted sense of new breeding communities that are infertile when crossed with the parent group. For example, all dogs form a single species because they are chemically capable of interbreeding. . . . (13)

Following publication of his book, Johnson began to make contact with others in the fields of science and philosophy who shared his skepticism of evolution and his belief that intelligent design explains the origin of biological life. As a result, Johnson, Dembski, biochemist Michel Behe, and Steve Meyer, a Ph.D. in the history and philosophy of science, gathered at Southern Methodist University in Dallas, Texas, in 1992 to discuss how to turn their beliefs into a movement. After several years of networking and connections, the conservative Seattle-based Discovery Institute and the supporters of the intelligent design movement formed the Center for Renewal of Science and Culture in 1996 to promote the idea of intelligent design (Menuge, 2004, 41).

Founded in 1990, the Discovery Institute refers to its mission as making "a positive vision of the future practical." It sees itself as "discover[ing] and promot[ing] ideas in the common sense tradition of representative government, the free market and individual liberty" ("About Discovery," 2010). Among the issues the organization addresses are regional transportation development, economics and technology policy, legal reform, and bioethics ("Media Backgrounder," 2004). The Discovery Institute's effort in promoting intelligent design initially was named the Center for Renewal of Science and Culture, but was eventually renamed simply the Center for Science and Culture. As of 2012 the Center had 40 fellows, scientists, and academic scholars who worked in publishing and public relations efforts on behalf of intelligent design.

Dembski writes that as a student at the University of Chicago in the 1980s, he believed in the traditional scientific conceptions. For example, he worked in a lab on chaos theory, which combines mathematics, physics, economics, and philosophy to investigate how minor changes in an environment can lead to widely diverging consequences. But he writes that he became increasing disillusioned with science until he began to develop the notion of intelligent design as a new scientific revolution (2004, 19).

EVIDENCE FOR INTELLIGENT DESIGN

To understand the intelligent design movement, it is essential to understand why proponents believe that the dominant approach to the origin of life, the theory of evolution, is a dehumanizing and outdated concept. Evolution, or as intelligent design advocates refer to it, Darwinism, is the idea that natural selection has led to the development of life over hundreds of millions of years. In a book for young Christians written by Dembski and Sean McDonald, for instance, Darwinism is criticized as an ideology, not a scientific theory, similar to Marxism and fascism (*Understanding Intelligent Design,* 2008).

The centerpiece of intelligent design is the idea of irreducible complexity, a system in nature that is composed of parts, the elimination of any of which will cause the system to fail (Behe, 1996, 39). Dembski writes that to be irreducible, a characteristic must be contingent, complex, or specified (2004, 5). Ruse summarizes Dembski's position: "Contingency is the idea that something has happened that cannot be ascribed to blind law. Being hanged for murder is contingent; falling to the ground when I jump off a stool is not." An example of complexity would be the series of prime numbers—those divisible only by themselves and 1, such as 1,3,5,7—all the way to 101. Specified means to match an independently given pattern (Ruse, 2007, 133–134).

Behe (1996) argues that irreducible complexity is a strong argument against Darwinian evolution, "Since natural selection can only choose items that are already working ... if a biological system cannot be produced gradually it would have to arise as an integrated unit, in one fell swoop, for natural design to have anything to work on" (39). As an example of irreducible complexity, Behe uses a mousetrap as an illustration. The mousetrap consists of several parts that work in tandem to kill a mouse. A mousetrap has a platform, a metal hammer that kills the mouse, a spring that sends the hammer toward the mouse, a catch that is released when the mouse steps on the platform, and a metal holding bar that holds the hammer in place until pressure is applied to the catch. Each part is essential for the proper operation of the mousetrap, and the alteration or elimination of any of those parts renders the mousetrap inoperable. Therefore, the mousetrap has irreducible complexity: It can't be changed or simplified. Other methods can be used to kill a mouse, such as a glue trap or a BB gun. But these other methods are neither a physical nor conceptual precursor of the mousetrap; you do not need to have a glue trap to build a mousetrap (42).

An example of irreducible complexity in biological systems is the human eye, Behe argues. Each part of the eye serves a discrete function. The retina perceives light, the lens focuses that light, and the muscles contract and focus the lens. Behe compares the eye to a stereo system, which is created by combining an amplifier, a CD player, a tape player, and speakers. He claims that "[e]ither Darwinian theory can account for the assembly of the speakers and amplifiers, or it can't" (38–39). For Behe and other intelligent design proponents, the eye is an example of irreducible complexity because every part is needed for the eye to work. If any part were missing, the eye could not function. Because all parts are necessary, it had to be created by a designer; if the eye had evolved, it would not have worked without all the parts, they contend.

Another example of irreducible complexity often cited by intelligent design proponents is a part of certain bacteria that is called the flagellum, a tube-like extension that allows the bacteria to "swim" through liquid. Behe's book describes the flagellum as a rotary propeller that is fueled by the flow of acid through a membrane (70–73). In a summary of his position, Behe describes how the flagellum's "outboard motor" is a long filament tail attached to a "drive shaft" through a hook region and a "motor." Proteins keep the "motor" stationary so the propeller will remain stable. If any part of the complex flagellum apparatus is missing, it will not work (2004, 353–354). Evolution supporters argue that the flagellum is the product of evolution. But Behe and intelligent design proponents contend that because of the complexity of the flagellum and the requirement of all parts to make a working flagellum, the flagellum could not have arisen through numerous, successive, slight modifications. That means, Behe says, that it is the product of intelligent design. "I have concluded the Darwinian processes are not promising explanations for many biological systems in the cell," he writes. "Instead . . . if one looks at the interactions of the flagellum . . . they look like they were designed—purposely designed by an intelligent agent" (355). Dembski argues, "If a creature looks like a dog, smells like a dog, barks like a dog and pants like a dog, the burden of evidence lies with the person who insists the creature isn't a dog. The same goes for incredibly intricate machines like the bacterial flagellum: the burden of evidence is on those who want to deny its design" (2004, 222).

Another example of irreducible complexity that intelligent design proponents cite is blood clotting. Behe writes that blood clotting at first appears to be a simple process; a cut bleeds for a while, and then the blood congeals, or clots, gradually stopping the bleeding. Instead of being a simple process, though, the combination of dozens of proteins creates a weblike structure in the blood where the cut occurred, trapping the red blood cells and creating a clot. Removal of any elements of the blood clotting system, Behe writes, would render the entire process unworkable, and therefore it is irreducibly complex (*Darwin's Black Box*, 74–97).

Another point made by intelligent design proponents critical of the theory of evolution lies with the fossil record. While evolution supporters say the existence of fossils proves that life has evolved over hundreds of millions of years, intelligent design supporters say two inconsistencies point to problems in evolutionary theory. First is the "gap" criticism, that the record of fossils contains gaps in which no fossils were discovered. Instead, they argue, a veritable explosion of fossils was deposited about 540 million years ago, during the Cambrian period. Before this fossil explosion, intelligent design proponents say, only a few simple invertebrates such as jellyfish, sponges and worms existed. Then, beginning in the Cambrian period, fossils of more complex animals, and eventually mammals, appeared. Strobel's book, *The Case for a Creator* (2004), recounts a conversation the author had about fossils with Jonathan Wells, a fellow at the Discovery Institute and the author of a leading intelligent design book, *Icons of Evolution* (2000). Asked to explain the timeline of this Cambrian explosion, Wells used the metaphor of a 100-yard football field. At one goal line are the first fossils, simple one-celled microscopic organisms. Past the 50-yard-line and down to the 16-yard-line on the other end of the field, the sponges, jellyfish, and worms appear. One stride farther down the field, the Cambrian explosion occurs with the appearance of most life forms (Strobel, 44).

A second criticism is the lack of fossils of transitional life forms. The textbook *Of Pandas and People* (1993) claims that once a life form has appeared in the fossil record and its taxonomy, or category, has been established, no fossils have been found that move from one category to another; according to this text, there have been no examples of fish turning into amphibians, or reptiles turning into birds. It argues that rather than illustrating changes into new or different species, the fossil record shows changes within species; for instance, animals that hunt get faster because the fastest animals have the greatest success in capturing game, eating well, and passing on their genes to their offspring. There is no doubt that average humans today are both taller and heavier than their counterparts 200 years ago. But intelligent design proponents say this intraspecies change over time does not address the origin of the species themselves. They believe in microevolution—small changes over time within species—but not macroevolution, which is the evolutionary transition among species.

In addition to the idea of irreducible complexity and a particular interpretation of the fossil record, Dembski and other intelligent design supporters also argue that the odds of life developing complex mutations from mutations are overwhelming and impossible. Dembski cites as an example a combination lock with the dial numbered from 0 to 39. If the dial has to be turned in alternate directions—for instance right, left, right—the number of possible combinations is 40x40x40, or 64,0000. A similar lock with numbers from 0 to 99 that has to be turned in five alternating directions has 10 billion possible combinations. Dembski writes that the odds of random mutation leading to a complex organism is 10^{150}, or 10 followed by 150 zeroes. And because, Dembski writes, computer and mathematical estimates state that anything less than 10^{120} is not likely the result of chance, then the odds state that random mutation into a complex form are impossible ("The Logical Underpinnings of Intelligent Design," 2004, 311–330).

One of the more unusual arguments on behalf of design has been made by theologian and former Biola University professor Jay W. Richards, who writes in *Intelligent Design 101* (2008) that solar eclipses prove design. According to Richards, the only place in the solar system where the moon can briefly block the sun entirely is on Earth. Other planets either have no moons or have moons that are too small to cause a total eclipse. Further, total eclipses are important for scientists because they have allowed for the discovery that the sun is made up of gases, the discovery of other stars, the development of Albert Einstein's Theory of Relativity (because Einstein was able to show that gravity can bend light waves), and other scientific breakthroughs. The alignment of the Earth, the moon, and the sun is perfectly designed, Richards contends, to lead to important scientific discoveries, leading to his theory that the universe is designed in a way that facilitates discovery (146–149).

As the examples of irreducible complexity, as well as the criticisms of the fossil record, and even Richard's eclipse argument indicate, proponents frame intelligent design as a scientific approach. Advocates such as Dembski and Meyer refer to the theory of intelligent design as a scientific research program based on observable phenomena. Dembski writes that because intelligent design is based on empirically detectible materials, it is a "full-fledged scientific theory" (*The Design Revolution*, 2004, 36–37). The Discovery Institute's website puts it this way:

> Intelligent design refers to a scientific research program as well as a community of scientists, philosophers and other scholars who seek evidence of design in nature. . . . Through the study and analysis of a system's components, a design theorist is able to determine whether various natural structures are the product of chance, natural law, intelligent design, or some combination thereof. Such research is conducted by observing the types of information produced when intelligent agents act. Scientists then seek to find objects which have those same types of informational properties which we commonly know come from intelligence. Intelligent design has applied these scientific methods to detect design in irreducibly complex biological structures, the complex and specified information content in DNA, the life-sustaining physical architecture of the universe, and the geologically rapid origin of biological diversity in the fossil record during the Cambrian explosion approximately 530 million years ago. ("Definition of Intelligent Design," 2010)

Another critic of evolution is author Wells, whose book *Icons of Evolution* (2000) attacks commonly held ideas about evolution, calling them hypotheses masquerading as theory. Among the "icons" he criticizes are the evolution of the horse, the development of humans from apes, and Darwin's Tree of Life. "If the icons of evolution are supposed to be our best evidence for Darwin's theory," he argues, "and all of them are false or misleading, what does that tell us about the theory? Is it science, or is it myth?" (8).

CRITICISMS OF INTELLIGENT DESIGN

Critics of intelligent design include many scientists who argue that intelligent design proponents misstate evidence, ignore that which does not fit their idea of a designer, and generally are theistic and unscientific. Evolution supporter Kenneth Miller, a professor of biology at Brown University, says that intelligent design advocates argue, for instance, that the modern horse did not evolve over 55 million years, as most scientists contend, moving from a small animal no larger than a house cat to today's modern horse (2008). Rather, Miller says, some intelligent design proponents believe each item in the "horse evolution tree" is actually a separate species: "What happened over time is the designer created a handful of little browsing species and then, as each one went extinct, he replaced it with a modified version. When those went extinct, he drafted another round of replacements, and then another, then another" (51). For Miller, intelligent design advocates believe the designer created all life forms, perhaps in a "puff of smoke" or by manipulating life in utero, allowing one species to give birth to a new life form (50).

Eugenie C. Scott, director of the National Center for Science Education, writes that the intelligent design perspective is based on non-scientific evidence that cannot be tested (2006). Wells' *Icons of Evolution*, she contends, is based on misleading or inaccurate information, and she joins most scientists in calling intelligent design a new version of creationism. Behe's notion of irreducibly complex biological mechanisms such as the flagellum and the human eye also has been roundly criticized as incomplete and misleading. Behe argues that all 30 proteins are needed to make the flagellum work, and in the absence of any of the 30 proteins, he says, the flagellum will not work any longer. But science has proved that many of the proteins have duties elsewhere in the bacterium cell. As Miller argues, "the flagellum isn't the custom-made, designed from scratch collection of closely matched elements that ID [intelligent design] likes to claim. It's much more like a collection of borrowed, copied, and jerry-rigged parts that have been cobbled together from the spare-parts bin of a cell. In short, it's exactly the sort of thing you'd expect from evolution" (61–62).

The functioning of the eyeball is another piece of evidence cited by intelligent design proponents. Every part of the eye is essential for it to work, the intelligent design advocates argue. But scientists say that the eye evolved from sightless bacteria over 500 billion years, and in fact there are examples in nature of intermediate eye development (Dennett, 2006, 36). Rather than proving the existence of a designer, they say, the eye actually is evidence of evolution. Furthermore, the eye doesn't function well at all; if it was created by a designer, evolution supporters say, it was designed quite poorly. The rods and cones, which detect light and color, go through a hole in the retina, creating a blind spot. In addition, the retina is inside out (37).

Intelligent design advocates argue that while there is evidence of microevolution, or change within species, the idea cannot be extended to macroevolution, the underlying idea behind evolution. For scientists, the question of microevolution and macroevolution are simply a question of units of analysis; when they look at changes within species, it is micro, and when they examine evidence of major biological shifts, its macro. It is all evolution (Matzke and Gross, 2006).

INTELLIGENT DESIGN AS SCIENCE OR NOT SCIENCE

One of the hallmarks of the scientific approach is the ability of scientists to have their original research published in peer-reviewed academic journals. Peer review is a system in which the authors (or author) of a study submit a manuscript to a journal; the editor sends the study, without the authors' names attached, to other scientists in the same field. They review the study and submit their recommendations to the editor as to whether the study should be published. Those reviews are forwarded to the authors without the reviewers' names attached. The editor may then reject the article; recommend publication; or recommend publication following certain changes to the manuscript. The authors may make the changes and resubmit the manuscript to the journal editor. It may (or may not) go through the review process again. This complicated system is designed to ensure that only the best, most rigorous science makes it into print.

Key questions for intelligent design advocates are whether their field has been subject to peer review and whether it qualifies as a legitimate science. Dembski contends that intelligent design has been peer reviewed and points to his own book *The Design Inference*, which he says went to three anonymous reviewers before it was published in 1998, although he never specified whom the reviewers were. The book *Debating Design* that he and Michael Ruse edited also went to seven anonymous reviewers before being published (*The Design Revolution*, 2004, 302). Further, the Discovery Institute lists a number of articles it refers to as peer reviewed, including articles published in *Dynamical Genetics, Proceedings of the Biological Society of Washington, International Journal of Design & Nature and Ecodynamics*, and *Physics of Life Reviews*, although there is no indication of the qualifications of those reviewers.

Most mainstream scientists are not convinced that intelligent design is a scientific approach, however. Intelligent design critic Peter J. Bowler, a biology historian, in his book *Monkey Trials and Gorilla Sermons* (2007), puts it bluntly: "The scientific community continues to treat ID with contempt" (213). The book *Critique of Intelligent Design*, edited by sociologists John Bellamy Foster, Brett Clark, and Richard York (2008), argues that intelligent design is religious irrationalism masquerading as science (25–26). Eugenie C. Scott's book *Evolution vs. Creationism* (2009) states, "There is, of course, no scientific way to test for the existence of the intelligent designer" (xii). The edited collection *Scientists Confront Intelligent Design and Creationism* (2007) calls intelligent design nothing but "creationism lite." Critics say the articles Dembski cites are not peer reviewed, meaning they have not been examined by other scientists and judged to be legitimate studies. But one of the harshest critiques of whether intelligent design is a scientific approach came in U.S. District Judge John Jones III's decision in *Kitzmiller v. Dover*, addressing the Dover, Pennsylvania, school board's attempt to require science teachers to question the theory of evolution and discuss intelligent design as an alternative theory of the origin of life. In a lengthy discussion on "Whether ID is Science," Jones writes that intelligent design meets none of the benchmarks that make something scientific, including its reliance on the untestable idea that a "supernatural" source is responsible for design. "While supernatural explanations may be important and have merit," he writes,

> they are not part of science. This self-imposed convention of science, which limits inquiry to testable, natural explanations about the natural world, is referred to by philosophers as "methodological naturalism" and is sometimes known as the scientific method. Methodological naturalism is a "ground rule" of science today which requires scientists to seek explanations in the world around us based upon what we can observe, test, replicate, and verify.... ID violates the centuries-old ground rules of science by invoking and permitting supernatural causation.... It is additionally important to note that ID has failed to gain acceptance in the scientific community, it has not generated peer-reviewed publications, nor has it been the subject of testing and research. (735)

Further, Jones writes, not a single mainstream scientific association accepts the idea that intelligent design is a scientific approach.

In addition, critics of intelligent design point out that science, including evolution, is used to make predictions about what fossils should be found in what rock and what genes can be mutated; intelligent design proponents cannot make predictions based on their "theory" because intelligent design is an ending point in analyzing biology and not a true science that advances through the discovery of new information (Poole, 2006, 52).

To design proponents, however, the fact that scientific groups discount intelligent design means simply that intelligent design investigators are challenging the status quo. In his argument on behalf of intelligent design, theologian John Frame contends that much of science is based on the unobservable, such as Einstein's theory of relativity, which came solely from his mind. "Many of his hypotheses have subsequently been verified by observation and experiment. Einstein was not himself an experimental scientist. But no one would deny that he was a scientist of the first order. The work of science, then, is not only observational and experimental, but also imaginative and logical. The scientist must use his imagination to determine significant hypotheses, and his logic to determine what it would take to verify or falsify these hypotheses and whether an experiment has, in fact, verified or falsified it" (2010).

Miller writes that intelligent design might never reach the goal of accepted science, that the idea of a "theistic" science that relies on God may never be accepted by the majority of people (198). But Dembski argues that whether most scientists accept intelligent design as a legitimate approach does not matter. The peer review process, he argues, is conservative. He claims that peer reviewers tend to look favorably only upon research that conforms to long-held views of what science is and what is an acceptable scientific approach. He argues that a research paper or article questioning the theory of evolution is akin to questioning Stalin in the Soviet Union, and he notes that many groundbreaking scientists never went through the peer review process, including Galileo, Newton, and—ironically—Darwin himself (*The Design Revolution*, 2004, 304–305). Further, Dembski and Jonathan Witt write in *Intelligent Design Uncensored* (2010) that intelligent design meets the scientific requirement that it be testable and able to be proved false. Behe's contention that the bacterial flagellum is irreducibly complex can be tested; if the flagellum is proved to be further reducible, then the theory is able to be proved false (143).

Another proponent of intelligent design, Dan Peterson, writes in *American Spectator* (2006) that critics who contend that intelligent design is not science are mainly interested in protecting "materialism" or "naturalism, or the belief in centrality of nature as opposed to the belief in the immaterial or spiritual world." Intelligent design is similar to other scientific approaches in that it attempts to find patterns in nature. One difference, however, is that it attempts to find patterns that are evidence of a designer's hand at work. It does not, however, necessarily attempt to establish a moral foundation, an explanation of who the designer is, or whether the designer has the attributes of the Christian concept of God.

Johnson, the law professor credited with jumpstarting the intelligent design movement with his 1991 book *Darwin on Trial*, claims nearly two decades later that the debate over evolution is not a debate about God, but rather a debate about science. "There is no question that we can look at *Archaeopteryx* and say it looks like something that could have been an intermediate between dinosaurs and birds," he writes. "There is also no question when we look at the bacterial flagellum, we can see that it closely resembles rotary engines found in cars and outboard motors designed by humans, which require all of their irreducibly complex parts in order to propel a bacterium. This," he believes, "constitutes a genuine scientific debate" (2008, 37).

CONCLUSION

Regardless of how scientists or judges view intelligent design, the idea has won some converts in the arena of public opinion, although it has a way to go. The Gallup Poll of public opinion, for instance, found in 2005 that 31 percent of the respondents believed intelligent design is definitely or probably true, while an almost equal percentage—32 percent—believed that intelligent design is probably or definitely false. The same question asked over the next four years found a consistent 35 percent to 40 percent believing that God had a hand in developing man over millions of years. The 2010 Virginia Commonwealth Life Sciences Survey of 1,001 adults found that 31 percent of the respondents believe scientists have serious doubts about evolution, and 43 percent believe that biological life developed over time from simple substances, but that God guided this process, which the survey analysis contends is consistent with the intelligent design approach.

The nation's political leaders have jumped into the fray, with some of them citing their own beliefs in questioning evolution. While running for the U.S. presidency in 1980, which he would ultimately win, Republican candidate Ronald Reagan stated that if schools teach evolution, they also should teach creationism. Another Republican candidate for president, Pat Buchanan, stated in 1995 that he did not believe he had descended from apes and criticized the teaching of "godless" evolution in the schools. Some state Republican Party groups even put a plank in their political platforms in the 1990s calling for creationism to be taught in the classroom (Numbers, 2007). Minnesota Governor Tim Pawlenty, who was described as a likely presidential candidate in the second decade of the twenty-first century, came out in favor of intelligent design during an appearance on NBC's "Meet the Press" in August 2008.

Whether intelligent design is creationism in disguise or a legitimate scientific approach, the intelligent design movement has generated tremendous controversy over the two-plus decades in which it has been on the national radar. One need only examine literally hundreds of books, magazine articles, and newspaper accounts to understand that there is no middle ground on the issue. The writers of these texts are either proponents of intelligent design, or they are unrelenting critics of the movement.

CHAPTER 2

Evolution

How did life develop? Was it the result of random mutations, the survival of the fittest, rapid changes brought on by changing environmental conditions, or a combination of these developments? Or was something else involved?

What is known today as the theory of evolution, sometimes called Darwinism by critics, emerged largely from the work of Charles Darwin, in particular his book *Origin of Species* (1859), as well as *The Descent of Man* (1871). In the 150 years since Darwin's death, his ideas have taken hold, and the theory of evolution is the only approach to the development of life that is accepted by mainstream scientists. The theory of evolution has attracted controversy, however. There are those who claim that the theory of evolution is nothing more than scientific orthodoxy, and that scientists are not interested in taking a new approach to the origin of life, specifically that of intelligent design.

This chapter will examine the development of the theory of evolution and the case its proponents make for it as the only generally accepted scientific approach to the origin of life, as well as what critics say about the theory of evolution.

DARWIN AND THE HISTORY OF EVOLUTION

Despite the attention paid to Darwin's role in the creation of the theory of evolution, he was not the first scientist to broach the subject. That distinction belongs to Jean-Baptiste Lamarck, a French soldier who, because of injury, trained as a botanist. After a decade of studying invertebrates, Lamarck theorized in *Philosophie zoologique* in 1809—the same year Darwin was born—that a change in the environment causes changes in what organisms living in that environment need, which in turn causes changes in their behavior. This process causes the failure to use some parts of the organism, which leads to structural changes over generations. He also contended in *Philosophie zoologique* and other books that all structural changes are inheritable. It was not until the early twentieth century that scientists conclusively determined that Lamarckism, as it became known, was flawed because characteristics added through an animal's efforts cannot be handed down through genetics (Bowler, 1989, 21). In the 1890s, for instance, German biologist August Weismann conducted an experiment with rats that disproved Lamarck's proposal. Weismann cut the tails off rats, then bred them; their offspring were born with tails. Weismann cut off the tails of this generation, bred the rats again, and again the next generation was born with tails intact. He did the same for 20 generations of rats, and the results were the same; changes in the biological samples could not be passed on to future generations (Scott, 2009, 84–85).

Darwin was born in England on February 12, 1809—the same day, incidentally, that future President Abraham Lincoln was born in America (Ruse, 2008, 1). Darwin was the son of a wealthy physician and financier, Dr. Robert Darwin. Being born into such a family, it was preordained that Charles would attend medical school and become a physician like his father. He attended the University of Edinburgh Medical School but found he was not interested in studying medicine and left without a degree. He was, however, interested in natural history. He came by this interest naturally; his grandfather, who died before Charles was born, was Dr. Erasmus Darwin, a physician whose interest in science and technology led him and others to form the Lunar Society, a scientific and technological society. In addition, Erasmus Darwin was an early evolutionist (Ruse, 2008, 1–2).

When Charles Darwin left medical school in Scotland he traveled to Cambridge, where he studied with both a geologist and a botanist while pursuing a general degree in Arts, which at the time was widely seen as a step toward becoming a minister in the Church of England.

At the end of his time at Cambridge, Darwin was given the chance to travel on a survey ship, the HMS *Beagle*. Between 1831 and 1836, the ship traveled from England to South America, around the southern tip and up the Pacific side of South America, then across the Pacific, around Australia and the southern tip of Africa, and finally back to England. But it was off the coast of Chile that Darwin made the discoveries that would change science.

On the Galapagos Islands, Darwin discovered a variety of finches. On each island, the finches had adapted a unique method of feeding, developed in isolation from the finches on the other islands. After the *Beagle* left the Galapagos, Darwin was told that the giant tortoises that were common to the islands also had unique characteristics from island to island. He returned to England with biological samples, including birds. After consulting with ornithologists, he learned that each variation of finch, for example, had to be considered as a separate species.

A religious man, Darwin was surprised by the idea of the vast variety within what had been considered to be a single species. As Bowler writes, "He was now faced with a dilemma: to preserve the traditional view that each true species was separately created, he would have to believe that God had performed a separate miracle for every finch and mockingbird species on these insignificant islands. He decided that this position reduced the creation hypothesis to absurdity" (Bowler, 2007, 87). Instead of the generally accepted view of God creating all life forms, Darwin's idea that natural selection—the adaptation of life forms to their physical surroundings—began to take shape. He published articles in a number of smaller journals espousing his growing idea, and in 1859, after two decades of exploration, consultation with other scientists, and thought, he published his groundbreaking work *Origin of Species*.

The central premise of his book, which eventually went through six editions with minor changes in each, was rather than being created by God as distinct forms of life, species evolve over long periods due to external environmental stimuli. Species that begin as one can evolve into different species if they are separated and encounter different environments. That explains how birds that started similarly grew into separate species.

A second argument Darwin made was that natural selection allows for slight mutations to become dominant features over time. Darwin's argument was that, for instance, animals that are slightly faster than their peers would have greater success at hunting prey and therefore a greater chance of surviving. Obviously, survival is critical to having offspring, and those offspring would also have the gene for being faster, making them more likely to survive than their slower peers. Over time, the "slow" gene would vanish, leaving only the animals with the "fast" gene. Through this process, over hundreds of thousands of years, animals adapt to their surroundings and evolve. As Darwin wrote in the introduction to *Origin of Species*:

> As many more individuals of each species are born than can possibly survive; and as, consequently, there is a frequently recurring struggle for existence, it follows that any being, if it vary however slightly in any manner profitable to itself, under the complex and sometimes varying conditions of life, will have a better chance of surviving, and thus be NATURALLY SELECTED. From the strong principle of inheritance, any selected variety will tend to propagate its new and modified form (5).

In Darwin's explanation of natural selection, as species evolve, the older forms of the same species become extinct:

> The theory of natural selection is grounded on the belief that each new variety and ultimately each new species, is produced and maintained by having some advantage over those with which it comes into competition; and the consequent extinction of less-favoured forms almost inevitably follows. It is the same with our domestic productions: when a new and slightly improved variety has been raised, it at first supplants the less improved varieties in the same neighbourhood; when much improved it is transported far and near, like our short-horn cattle, and takes the place of other breeds in other countries. Thus the appearance of new forms and the disappearance of old forms, both those naturally and artificially produced, are bound together. In flourishing groups, the number of new specific forms which have been produced within a given time has at some periods probably been greater than the number of the old specific forms which have been exterminated; but we know that species have not gone on indefinitely increasing, at least during the later geological epochs, so that, looking to later times, we may believe that the production of new forms has caused the extinction of about the same number of old forms. (320)

Darwin saw the evolutionary process as a "tree of life," with the extinct species forming the roots, the main groups of organisms as branches, and life today as the leaves and buds at the end of the branches. The tree was organic—not ordained by a supreme being—and his depiction of this tree was the only diagram he included in *Origin of Species* (Browne, 2006, 72–24).

REACTION TO DARWIN

Predictably, Darwin's theory of evolution through natural selection was not well received by either the Church of England or by those who believed in the status quo idea of creation by an omnipresent

God. In one well-publicized reaction, one of Darwin's supporters, Thomas Henry Huxley, clashed in a debate with Samuel Wilberforce, the Bishop of Oxford. The debate occurred in 1860 at a meeting of the British Association for the Advancement of Science. Although Darwin had barely broached the subject of the evolution of humans at this point, Huxley and Wilberforce argued forcefully over the question of whether humans had evolved from a lower life form or were created by God as described in the book of Genesis. Legend has it that when Wilberforce asked Huxley whether he was descended from apes on his grandmother's or grandfather's side, Huxley told the Bishop he would rather be descended from an ape than from a Bishop, although Bowler (1989) writes that Huxley probably said he would rather be descended from an ape than from a man who used his talent to attack a theory he misunderstood. "Whatever his actual words," notes Bowler, "the result was an uproar in the audience—ladies fainted and Robert Fitzroy (Darwin's old captain on the Beagle) stalked around waving a Bible. On one point the legend is clear, however; Huxley had carried the day, and Wilberforce was crushed" (107).

Despite criticism from religious leaders, who objected to any theory that did not include God creating life, Darwin and his supporters continued to publish their ideas. Huxley published a book in 1863 titled *Evidence as to Man's Place in Nature*, a follow-up to *Origin of Species* in which Huxley argued that the end result of evolution had to be movement to higher beings. While Darwin conceived of evolution as a slow process, Huxley saw it as moving much more quickly, with evolutionary "jumps." In addition, while Darwin had expressly avoided the topic, Huxley wrote about the evolution of humans, using fossil evidence to argue that man had descended from lower forms of life. Huxley's book was the first attempt to apply the theory of evolution directly to humans. Huxley, who had dissected primates while preparing his book, found, for instance, that both humans and apes had a hippocampus and that there was more similarity between apes and humans than between humans and other primates such as lemurs (Gould, 1977, 49).

Darwin followed in 1871 with *The Descent of Man*, in which he argued that a close relationship exists between humans and apes, both physically and mentally. He wrote that humans developed in Africa and that primitive humans had moved from trees to the plains as they adapted to changing environmental conditions (Bowler, 1989, 126). Predictably, many leaders of organized religion had difficulty accepting Darwin's theory of man evolving from a lower form of life, rather than being the end result of God's creation of life. Much of the opposition to Darwin came from religious leaders in the United

States, where strongly conservative religious traditions led many to argue that the Genesis description of how God created life meant they had to reject evolution as a Godless attempt to mislead the faithful. Press reaction to Darwin and his theories was at times critical, with his approach being "framed," or depicted, as a conflict between science and religion, a theme that continues to this day. Many press accounts framed Darwin's theory as atheistic science opposed to the Christian belief that God had created all plants and animals (Caudill, 1997, 138).

In his later years, Darwin suffered from ill health; he became a virtual recluse between the publication of *The Descent of Man* in 1871 and his death in 1882, although he continued to publish minor tracts on plants, including his final book, *The Formation of Vegetable Mould through the Action of Worms*. But even in death, Darwin remained controversial. During his life, Darwin had moved away from belief in traditional orthodoxy and became an agnostic. Despite no evidence, some critics of Darwin claim that he had a deathbed conversion and stated his belief in God (Caudill, 1997, 55).

THE THEORY OF EVOLUTION EXPANDED

While some of Darwin's ideas have been disproved over the years, his general thesis that evolution occurred through natural selection "was a sophisticated theory backed at every point by his copious data—one that has inspired scientists ever since," and "initiated a revolution in how we understand nature" (Foster, Clark, and York, 2008, 137–138). Darwin's *Origin of Species* has been hailed as one of the most important books ever written (Quemman, 2006, 174). Even critics such as intelligent design advocate Michael Behe (1996) note the elegant simplicity of Darwin's ideas. Darwin's revolutionary ideas changed the way biologists, geologists, and other scientists view the natural world. In this sense, it is hard to overestimate the importance of Darwin's theories.

Some intelligent design proponents argue that Darwin's "materialistic" view—that life evolves from material rather than being created by a supreme being or other intelligent designer—has changed little since the 1800s. Leading intelligent design proponent William Dembski, for instance, claims in *The Design Revolution* (2004) that "no significant details have been added since the time of Darwin" (270). However, scientists who support the study of evolution call this notion of a stagnant field "absurd" (Foster, Clark, and York, 128). In her book *Darwin's Origin of Species: A Biography* (2006), Janet Browne notes that beyond Darwin's scientific advances, the publication of *Origin of Species* allowed liberal theologians to accept the idea that nothing in the Bible ruled out the possibility of evolution over millions of years.

In the years that followed Darwin's publications the advances in the study of evolution moved from the observable differences in live biological specimens to the discovery and examination of more and more fossils. Although knowledge of the existence of fossils predated Darwin, it was not until the late nineteenth and early twentieth centuries that the study of fossils became a discipline in its own right. Some of the archaeologists who worked to uncover these fossils contended that rather than proving Darwin's "tree of life," the fossil record showed that evolution moved in a straight line. Among these were the archaeologists Elmer Eimer, Alpheus Hyatt, and Edward Drinker Cope. In particular, Henry Fairfield Osborn, president of the American Museum of Natural History from 1908 to 1933, argued as early as 1898 that the fossil record proved that each group of organisms developed rapidly at first, then stabilized into steady lines, which then developed further (Browne, 121–122). He also promoted the idea of rare but occasional rapid divergent revolution following mass extensions, what he called "adaptive radiation" (Bowler, 1989, 146).

In *The Panda's Thumb: More Reflections on Natural History* (1980), the noted scientist Stephen Jay Gould discusses Belgian paleontologist Louis Dollo, who developed an important modification to Darwin's idea of how organisms evolve. Dollo contended that evolution was irreversible—that two separate lines of organisms may evolve into similar beings, but because the organisms have so many complex parts, the chance of two organisms evolving twice toward the same result is impossible. Two organisms may evolve in such a way that they appear to be similar, but that similarity will be superficial only (38–39). In particular, Gould was criticizing Arthur Koestler, who argued for the similarities in skulls between wolves and the "Tasmanian wolf," a marsupial more related to kangaroos and koalas that wolves. Koestler contended that the skulls of the two species were identical, thus disproving the theory of evolution, because for two distinct species to have evolved in exactly the same way is a statistical impossibility. But Gould argues that the wolf and the Tasmanian wolf have similar skulls because they both hunt for live food: "Random variation may be the raw material of change, but natural selection builds good design by rejecting most variants while accepting and accumulating the few that improve adaptation to local environments" (40).

Further adding to the study of evolution was the work of Gregor Mendel, an Austrian monk who lived from 1822 to 1884. During his lifetime, Mendel, who lived and worked in obscurity, discovered the basic ideas of genetics and heredity while studying pea plants. While he presented some of his research at scientific meetings and published some of it in 1865, his work was largely ignored during his life. Although Mendel was a contemporary of Darwin's, there is no evidence that Darwin knew of Mendel's work or that Mendel's work influenced Darwin. In fact, after Mendel's death, his papers were burned. Even though Mendel's work predated Darwin's *Origin of Species*, it was not until the 1900s that scientists looking into how traits are passed on to offspring finally began to recognize the groundbreaking discoveries contained in Mendel's work. Further research in the field of genetics uncovered the existence of recessive genes: traits passed from parents to offspring that may lie dormant for generations before finally emerging in an organism. The study of genetics in the early twentieth century refined Darwin's ideas of how the variation, adaptation, and selection of species occurred, showing that mutations in the genetic code—happy accidents—led to changes in species (Browne, 134–136).

The famous Scopes "monkey" trial in Tennessee in 1925, described in detail elsewhere in this book, resulted from a renewed interest in Darwin's work and the theory of evolution. That Tennessee legislators felt the need to enact a law banning the discussion of evolution of humans from lower orders speaks not only to the fear of evolution overtaking the conservative religious viewpoint of God's creation of mankind, but also to the general renewed scientific interest in Darwin's theory.

It was not until the 1940s, however, that the fields of evolution and genetics began to merge into a new scientific approach to the origin of life. A group of scientists, including Huxley's grandson, Julian Huxley, started looking at genetic variability. Mutations in chromosomes are a way in which attributes can be passed on to offspring. These mutations, Huxley and his colleagues discovered, presented a wide array of variations; therefore, in a given population, there are enough differences for natural selection to be viable (Browne, 138).

A University of Chicago geneticist, Sewall Wright, is credited by many with reinvigorating Darwin's idea of natural selection, along with scientists Ronald A. Fisher, J. B. S. Haldane, and others. Not only was Wright's work with guinea pigs and rats groundbreaking, but he was also able to describe his research in language that was easy for other scientists to understand (Browne, 139). Wright and his contemporaries became convinced that natural selection was the only logical explanation for the origin of species (Bowler, 1989, 156). The 1953 discovery by James Watson and Francis Crick of the structure of DNA and its the hereditary nature provided further evidence supporting Darwin's theory of natural selection. As Scott (2007) writes, "It is safe to say that by the mid-twentieth century, mainstream science in both

Europe and the United States was unanimous in accepting not only the common ancestry of living things but also natural selection as the main—although not the only—force bringing about evolution" (85).

Evolution depends on mutations in genes, for, as Stephen Jay Gould notes, "natural selection cannot operate without a large set of choices." Individual organisms "are the units of selection." Those individual organisms cannot evolve; they simply live out their lives without adapting. Evolution occurs in species, or groups of organisms and their offspring (1980, 85). Rather than the slow development that many early evolution supporters theorized, scientists now believe that evolution is punctuated by rapid changes—rapid in terms of hundreds of millions of years. These rapid changes occur during major transitions, such as when fish developed into amphibians or when reptiles evolved into birds (Dodson and Dodson, 1985, 544).

THE FOSSIL RECORD

One issue for scientists has been to estimate the age of both the Earth and of life on Earth. While some creationists place the age of the Earth in the thousands of years, as they say the Book of Genesis depicts, most scientists believe that the best estimate of the Earth's age is about 4.5 billion years (Gould, 1977, 150). Critics of evolution (and supporters of intelligent design) contend that the fossil record points to a Cambrian "explosion" of species that occurred about 540 million years ago. Scientists, however, say that while the huge amount of fossils found in rocks dating to 540 million years is interesting, it does not prove that species were not around before that period. Scott, for example, argues that the Cambrian fossils were deposited during a 15-million to 20-million year span, and there are plenty of fossils from before that period. Trace fossils date back to 60 million years before the Cambrian fossils, approximately the same amount of time that has passed since the dinosaurs became extinct, providing more than enough time for evolution to occur (197–198). In his book *Darwin: A Very Short Introduction* (1982), Jonathan Howard also points out the long fossil history, writing, "The fossil record might have stopped dead at 4004 B.C. [the year some creationists put on the age of the Earth], but it in fact goes back about three hundred thousand years" (104).

In addition, scientists now believe evolution to be a slow process punctuated by periods of rapid change. This concept of rapid evolutionary change would explain the lack of transitional fossils; the changes may have been so rapid that there were fewer fossils left behind. For instance, scientists believe that modern-day birds evolved from reptiles, as evidenced by their hollow bone structure; the fact that their claw structure is similar to dinosaur claws; and the fact that that archaeopteryx had feathers. In terms of geological time, the evolution of birds from reptiles would have taken place fairly rapidly.

EVOLUTION OF HUMANS

Nothing fuels criticism of both Darwin and the theory of evolution more than the idea that humans and apes developed from a common ancestor. Critics refer to the idea that man descended from "monkeys" and argue that if man had developed from apes, there is nothing inherently good or moral about mankind; descent from apes would mean that humankind is an "accident" that is the result of chance. As the intelligent design-oriented textbook *Of Pandas and People* (1993) puts it, Darwin's *The Descent of Man* theorized that humans evolved from "some creature" (107).

While there have been countless fossils found of everything from prehistoric invertebrates to dinosaurs to the ancient precursors of modern-day animals such as horses and elephants, the fossil record of the development of humans has been much sketchier, leading to another criticism of evolution by intelligent design proponents. While there have been plenty of fossils of humanoids, beginning in 1890 with the discovery of *homo erectus*, which was considered to be the ancestor from whom *homo sapiens* evolved, discoveries in the 1920s of more ancient remains in Africa that scientists named *Australopithecus* pushed scientists' understanding of humans' predecessor even further (Bowler, 1989, 157). In 1974, archaeologists discovered Lucy, an *Australopithecus afarensis* fossil in Ethiopia, a discovery that proved that primates had been bipedal for at least three million years.

The "discovery" of the so-called Piltdown Man—later found to be a fraud—still causes some critics to question the reliability of evolutionary science. "Discovered" in Piltdown in South England 1912, Piltdown Man consisted of a modern-looking, human-type skull with an ape-like lower jaw. At the time, the remains were hailed as the "missing link" between ancient ape-like humanoids and modern humans. Few at the time believed it to be a hoax, although some did suspect that two separate items—a human skull and an ape's jaw—had been combined. For 30 years, Piltdown Man held the public's and scientists' interest. Eventually, scientists who examined the skull and jaw determined they had been artificially stained to appear to be from the same site, and that flints and bones allegedly found with the remains had been worked on with modern tools rather than ancient flints. The skull was human, but the jaw was from an orangutan. No one confessed to the fraud, though several books and articles have theorized who could have been to blame (Gould, 1980, 108–124). Regardless of who was responsible for the hoax, the scientific community in general and advocates of the evolution of humans from lower orders suffered an embarrassing black eye. Critics of evolution still use "Piltdown Man" as an example of why evolution is fundamentally flawed.

At the dawn of the twenty-first century, however, scientists began to unlock humans' genetic code, and their research revealed genetic similarities between modern humans and Neanderthals, the stockier, low-browed "cavemen" from Europe. Fossil evidence shows that Neanderthals, who began dying out 30,000 years ago, coexisted with modern humans, who had been moving north from Africa, 40,000 years. One study suggests that humans and Neanderthals shared a common ancestor and, in fact, that some humans and Neanderthals mated, mixing their genetic material (Kaufman, 2007). Studies show that 99.5 percent of the genes humans have today are shared with the genes of Neanderthals, which scientists say is further evidence that humans developed from prehistoric ancestors.

Discoveries of humanoid fossils continue. In 2008, researchers in South Africa found fossil remains of an early human-type boy and girl from 1.95 million years ago. The fossils, which the researchers named *Australopithecus sediba* (meaning southern ape wellspring), had a mixture of early human and ape features, including long orangutan-like forearms and a small skull, along with a flat face and long-striding hips (Veragno, 2010).

Those who argue against the evolution of humans from lower forms often couch their opposition in terms of morality and the essence of what it means to be human. The Catholic Church, for instance, teaches that only humans have a soul and that the soul was created by God. Apes do not have souls, according to Catholic teachings, and therefore humankind could not have evolved from non-human ancestors (Trigilio and Brighenti, 2007, 12). If humankind developed from an ape-like ancestor, critics argue, then people are nothing more than brutes who have no higher, divine calling. Others argue that if humans evolved from lower forms, that makes humans just another mammal with no moral center. But evolution supporters argue that humankind's evolution, while the result of chance, does not mean that humans have not developed intrinsic moral beliefs. Indeed, some people, called evolutionary creationists, contend that nothing in the Bible proves that evolution has not occurred, and that God may have had a hand in creating the primordial soup from which life evolved.

CRITICISMS OF EVOLUTION

Critics of evolution also point to what they call gaps in the fossil record, or an absence of transitional forms of species proving that life evolved from lower forms. Instead, intelligent design supporters argue, all species originated as they are today, with minor intra-species changes over time but no major shifts. In summarizing this viewpoint, Scott's *Evolution v. Creationism* (2009) argued that if it were possible to find single-celled fossils in Precambrian rock—rock that was created before the "Cambrian explosion"—then museums should be filled with the fossils of transitional forms between simple creatures such as bacteria and algae and more advanced vertebrates; however, intelligent design proponents claim, no such creatures have been found. Moreover, scientists note that fossils are created only in particular geologic conditions; not all living things leave fossil evidence behind. As Bowler writes in *Monkey Trials and Gorilla Sermons*, "there is no reason to expect that the fossil record will be a continuous record of all changes that take place in the history of life. There will inevitably be gaps, some of which we will fill in with future discoveries, but some of which will be permanent because the missing species were never fossilized" (2007, 100).

Intelligent design proponents who contend that alternatives to evolution should be taught in the public schools argue that they are merely suggesting that schools "teach the controversy," that evolution has not been proved and is merely a theory that is open to question. Mainstream scientists, however, argue that the theory of evolution has indeed been proven, and is the only accepted and scientifically backed approach to how life originated. Groups such as the National Academy of Sciences, the National Science Foundation, the National Science Teachers Association, and the American Association for the Advancement of Science all dismiss the notion that intelligent design is a legitimate field of study and support evolution as the only viable approach to teaching science. It is difficult, *The New York Times* reported in 1982, "to find any prominent biologist who disputes the theory of evolution" (Boffey, 1982). Even the Catholic Church, which has had a history of opposing evolution, has accepted the validity of evolution. Pope John Paul II issued a statement in 1996 acknowledging that evolution is "more than just a theory." The Pope said the human soul is of immediate divine creation, but he also noted that scientific evidence seemed to validate the theory of evolution through natural selection (Applebome, 1996, A-12). John Paul II's successor, Pope Benedict XVI, however, had a different take on evolution, writing in 2007 that while he did not reject evolutionary theory, he believed that science cannot fully explain the mystery of creation (Fisher, 2007, A-6).

Critics of the theory of evolution also cite the evolution of the idea itself. Evolution as a metaphor was adopted by social critics (that is, the notion of the survival of the fittest was applied to human endeavors). Herbert Spencer, a nineteenth-century philosopher, popularized the idea of social Darwinism, primarily the idea that only the strong survive. In fact, Spencer coined a phrase often incorrectly attributed to Darwin: "survival of the fittest." Spencer used social Darwinism to argue that white Europeans were the fittest, and that they had an inherent right to rule over other races. Business leaders also leapt on the social Darwinism bandwagon, arguing that the free market system meant that some would fail and only the strongest would survive, a notion particularly held by nineteenth- and early twentieth-century industrial leaders such as oil baron and monopolist John D. Rockefeller (Browne, 105–106). Critics of evolution also note that Karl Marx was fascinated by Darwin's theory and sent Darwin an inscribed copy of his book *Das Kapital*, although Browne writes that it was because Marx had confused of Darwin's son-in-law, who endorsed the secular approach to evolution, with Darwin himself (101). Marx's adoption of the evolution metaphor was in the idea of the economic formation of society as a process of natural history and in the idea of class struggle. Critics of evolution also cite Sigmund Freud's embrace of Darwinism as evidence of its corrupting nature. As the "Wedge Document" attributed to and acknowledged by intelligent design proponents claimed,

> Debunking the traditional concepts of both God and man, thinkers such as Charles Darwin, Karl Marx, and Sigmund Freud portrayed humans not as moral and spiritual beings, but as animals or machines who inhabited a universe ruled purely by impersonal forces and whose behavior and very thoughts were dictated by the unbending forces of biology, chemistry and nature. (Hume, 2007, 75)

Evolution supporters, however, do not contend that Darwin's ideas should or could be applied to the business world, Marxism, psychoanalysis, or racial superiority. In fact, critics of social Darwinism argue that the philosophy was behind both the eugenics movement, which held that social ills could be eliminated by improving the genetics of the human population, and the Nazi movement's attempt to eliminate Jews to maintain the purity of the Aryan race (Caudill, 1997, 66). Critics of social Darwinism call it a deliberate misstatement of the theory of evolution in an attempt to explain the inferiority of other people. Edward Caudill, in his examination of how Darwin's theory was misused, writes that social Darwinism was a popular idea because it could be adapted by anyone to fit a preconceived notion of who was the fittest and deserved to survive, and who was not fit and did not deserve to survive. In addition, he contends that social Darwinism appealed to the middle and upper classes because of its roots in scientific theory (1997, xiii).

One of the centerpieces of the intelligent design movement is Michael Behe's contention of "irreducible complexity" in biological systems, which he argues is proof that life was created by an intelligent designer (1996). In particular, Behe and many other intelligent design advocates cite the bacterium flagellum, which Behe metaphorically compares to the outboard motor on a boat (as discussed in the previous chapter). For Behe and his supporters, if any part of the flagellum is removed, it no longer would work, thus proving its irreducible complexity. Casey Luskin, another intelligent design proponent, summarizes the flagellum argument, writing, "it fails to function properly if one mutates or removes any of its approximately thirty to fifty genes" (2008, 86). Evolution supporters, however, contend that Behe is wrong in assuming the flagellum was created as it exists today. Rather, they argue, the flagellum is the result of an evolutionary process. They also note that Behe's book did not undergo the rigorous scientific peer review process. Further, during the Dover, Pennsylvania, trial on whether intelligent design could be taught in the public schools (*Kitzmiller v. Dover*, 2005), Brown University biologist and author Kenneth Miller testified that the flagellum was not, in fact, irreducibly complex. He also criticized Behe's metaphor that the flagellum is like a mousetrap; remove any part, Behe writes, and the mousetrap no longer works. Miller replaced the parts of a mousetrap with items such as a tie clasp, a key chain, a money clip and a paperweight, and the mousetrap still worked, which Miller says help to disprove the idea of irreducible complexity (Hume, 2007, 263–264).

Evolution also has been questioned by conservative political commentators, sometimes in odd fashion. Radio personality Glenn Beck, for instance, questioned evolution on his show in October 2010 because, he said, "If humans had evolved from apes, I haven't seen a half-monkey, half-person yet. Did evolution just stop?" (McGreal, 2010). In a similar vein, Christine O'Donnell, who would later run for U.S. Senate in Delaware, said during an appearance on a television show 12 years earlier that she considered evolution to be a myth, asking if evolution were true, "[W]hy aren't monkeys still evolving into humans?" (Flam, 2010, D1). Of course, evolution works over time; it does not occur instantaneously so it can be observed in action, and it does not create half-creatures.

Another criticism of evolution, particularly by non-scientists who respond to public opinion surveys, as well as by some public school officials who favor teaching either intelligent design or creationism as an alternative, is that evolution is just a "theory" that has not been proved and therefore should not be taught as fact. Evolution proponents counter that the term "theory" is being misunderstood. They cite the theory of gravity, which no one doubts exists, as representative of the way that the term "theory" is used by scientists. Scott, in her 2009 book, compares the way the public views the scientific terminology with the way scientists use the same terms. The public thinks facts are most important, followed in descending order by scientific laws, theories, and hypotheses. Instead, scientists consider theories to be most important, followed in descending order by laws, hypotheses, and facts. For scientists, a theory is not a guess, but instead is a comprehensive explanation of some aspect of life that is supported by a large body of evidence. Unlike a hunch, a theory has been thoroughly documented by accumulated evidence (11–14).

CONCLUSION

While intelligent design advocates characterize the theory of evolution as flawed and open to debate, for most mainstream scientists and scientific organizations there is no doubt that the validity of the theory of evolution has been established. The National Science Board, the governing body of the federal government's National Science Foundation, states that evolution is established science. As it said in a statement about the Kansas Board of Education's decision to remove evolution as a topic required in the state's public schools (a position from which the Board of Education later backed down), evolution "is a well-documented process—and the rich scientific debate about its precise nature will continue to contribute to our knowledge base" ("National Science Board Statement," 1999).

It is certain that everything about how life originated is not known; there are vast unanswered—and perhaps unanswerable—questions about how species began. Evolution supporters acknowledge that they do not have all of the answers, prompting them to do further research. Intelligent design supporters say these unanswered questions can be addressed through evidence of an intelligent designer. Evolution supporters say these same unanswered questions will be addressed through mainstream scientific research—that they will be answered through science, not belief in a "designer."

As Miller writes in *Only a Theory* (2008), evolution means that humans have a history that we share with every other living thing on Earth and that the same DNA sequences that control cell division in microscopic creatures also control cell division in humans (220).

It seems certain that intelligent design proponents will never accept the theory of evolution as the sole explanation of how life developed. Intelligent design proponents will continue to call the theory of evolution into question, and evolution supporters will continue to call intelligent design an unscientific new form of creationism.

CHAPTER 3

Intelligent Design in the Public Schools

Throughout the United States, local school districts and statewide education boards have been grappling with challenges to the theory of evolution and propositions to teach the concept of intelligent design, the idea that certain components of life are too complex to have been created by chance and that a "designer" had to have had a hand in the creation of life (Gunn, 2004). In particular, proponents of intelligent design challenge the teaching of evolution in public schools because they say evolution is a mere theory, while critics contend that intelligent design is an attempt to inject religious beliefs into the schools. Generally, courts have ruled that intelligent design is religious in nature because of the belief in a "designer" who is the Christian God. Intelligent design proponents, however, argue that their approach to the origin of life is agnostic, and that they do not say who or what the designer was.

This chapter will examine how the intelligent design-evolution debate has played out in the public schools, including the arguments both for and against teaching intelligent design in public schools; in the process, it will provide an overview of the numerous court cases that have been decided on the issue of evolution and the schools.

THE BEGINNINGS OF INTELLIGENT DESIGN IN THE SCHOOLS

One of the major reasons the concept of intelligent design made its way into the public schools was the publication of the textbook *Of Pandas and People* in 1989 (Davis, Kenyon, and Thaxton). The book casts doubts on the theory of evolution and proposes that intelligent design be taught as a reasonable approach to the origin of life. The book, despite being roundly criticized both by science teachers and book reviewers, became popular among members of the public who opposed the theory of evolution being taught in the schools (Johnson, 1991, 226). In Alabama, 11,000 parents signed a petition calling for *Of Pandas and People* to be adopted (227).

Of Pandas and People, which was published in a somewhat expanded second edition in 1993 with authors Davis and Thaxton, draws an analogy between evolution and the discredited 1600s belief in spontaneous generation, the idea that animals arose on their own from nonliving material. The book acknowledges that some things in nature are the result of natural causes, such as clouds, but it says that other things are the result of a designer, similar to words written in sand on the beach:

> When we find "John Loves Mary" written in the sand, we assume it resulted from an intelligent cause. Experience is the basis for science as well. When we find a complex message coded into the nucleus of a cell, it is reasonable to draw the same conclusion. . . . [W]hen scientists probed the nucleus of a cell, they stumbled upon a phenomenon akin to finding "John Loves Mary" in the sand. (1993, 7)

Of Pandas and People argues that saying DNA and protein found in a cell nucleus are the result of natural causes is like arguing that "John Loves Mary" arose from natural wave action.

The book also attempts to cast doubt on evolution supporters' arguments that natural selection leads to the development of new traits in a species—for example, that a giraffe's neck is long because giraffes eat leaves on trees, and those with long necks were more likely to survive and pass on their genes to offspring than were those with slightly shorter necks; therefore, giraffes have developed longer necks over time. *Of Pandas and People* argues that the giraffe's neck is combined with a coordinated blood pressure control system (because it raises its neck up to eat leaves, then lowers its head to eat grass or drink water). "In short," the authors write, "the giraffe represents not a mere collection of individual traits but a package of interrelated adaptations. It is put together according to an overall design that integrates all parts into a single pattern" (13).

The idea of teaching intelligent design in the public school science curriculum, as advocated by *Of Pandas and People*, has led to several high-profile, well-publicized court cases. The Discovery Institute's

Center for Science and Culture, while the leading proponent of intelligent design, rarely intervened in these court challenges. Instead, the Discovery Institute has provided moral support for efforts to require the teaching of intelligent design in the public schools while not actually providing witnesses in court. The Institute's website states the reason for this approach:

> All of the major pro-intelligent design organizations oppose any efforts to require the teaching of intelligent design by school districts or state boards of education. The mainstream ID movement agrees that attempts to mandate teaching about intelligent design only politicize the theory and will hinder fair and open discussion of the merits of the theory among scientists and within the scientific community. Instead of mandating intelligent design, the major pro-ID organizations seek to increase the coverage of evolution in textbooks by teaching students about both scientific strengths and weaknesses of evolution. Most school districts today teach only a one-sided version of evolution which presents only the facts which supposedly support the theory. But most pro-ID organizations think evolution should be taught as a scientific theory that is open to critical scrutiny, not as a sacred dogma that can't be questioned.

Rather that advocating the direct teaching of intelligent design in the public schools, William Dembski of the Discovery Institute and some other intelligent design proponents instead argue that students in science classes should be taught simply to question the theory of evolution—what advocates call "teaching the controversy." By arguing that the theory of evolution has inherent flaws and should not be accepted as dogma, these intelligent design advocates say teachers should be raising questions that students can find answers to outside of class. As Dembski and Jonathan Witt (2010) write, "[E]ncourage states to teach both the strengths and weaknesses of modern evolutionary theory. . . . Give students the evidence for and against evolution—all of which can be found in the mainstream, peer-reviewed scientific literature—and allow students to evaluate the data" (147). Some intelligent design advocates would go further. David DeWolfe, Stephen C. Meyer, and Mark E. DeForest contend in their chapter in *Darwinism, Design and Public Education* (2003) that allowing teachers to raise doubts about evolution inevitably would lead to students asking for an explanation of what theory would answer their questions. "Good science cannot require teachers to refuse to answer such a question," they write, arguing that teachers should be free to discuss intelligent design as an alternative to evolution (82).

It is clear that many members of the public agree with the idea that the theory of evolution has flaws that should be discussed. Surveys show that a majority of people in the United States support the questioning of evolution in the schools, and a 2005 survey of 2,000 people by the Pew Forum on Religious Life and the Pew Research Center for the People and the Press found that 42 percent of the respondents were creationists and that 64 percent of respondents said they were open to the idea of teaching creationism in the schools (Lester, 2005). But while many people would approve of school policies requiring science teachers to question the theory of evolution in science classrooms, judges across the country have ruled repeatedly that teaching intelligent design is inserting a religious viewpoint in the classroom in violation of the First Amendment's prohibition against the establishment of an official national religion.

THE DOVER, PENNSYLVANIA, CASE

The highest-profile and perhaps most contentious fight over evolution and intelligent design occurred in the Pennsylvania community of Dover, a city of about 90,000 residents approximately 25 miles from the state capital of Harrisburg, where the local school board attempted to require school system science teachers to give "equal time" to evolution and intelligent design.

The Dover dispute began in 2004, when a majority of the nine-member Board of Education rejected a high school science textbook recommended by the school system's science teachers; the board members objected because the textbook discussed Darwin and the theory of evolution. The books were eventually ordered, but the board ordered teachers to allow an administrator to read a statement to their students casting evolution as an unproven theory and stating that intelligent design is another legitimate approach to the origins of life. The statement said:

> The Pennsylvania Academic Standards require students to learn about Darwin's theory of evolution and eventually take a standardized test of which evolution is part. Because Darwin's theory is a theory, it continues to be tested as new evidence is discovered. The theory is not a fact. Gaps in the theory exist for which there is no evidence. A theory is defined as a well-tested explanation that unifies a broad range of observations. Intelligent Design is an explanation of the origin of life that differs from Darwin's view. The reference book *Of Pandas and People* is available for students who might be interested in gaining an understanding of what Intelligent Design actually involves. With respect to any theory,

students are encouraged to keep an open mind. The school leaves the discussion of the origin of life to individual students and their families. As a standards-driven district, class instruction focuses upon preparing students to achieve proficiency on standards-based assessments. (*Kitzmiller v. Dover*, 2005)

In addition, 50 copies of the textbook *Of Pandas and People* were placed in the school library.

The six school board members who voted for the new policy and their supporters contended that a balanced approach to the origin of life necessitated the inclusion of intelligent design as an alternative to the unproved theory of evolution. Attorney Richard Thompson from the conservative Christian-based Thomas More Law Center in Ann Arbor, Michigan, who agreed to represent the board, said inclusion of intelligent design made logical sense: "When you look at cell structure and see the intricacies of the cell, you come to the conclusion that it doesn't happen by natural selection" (Badkhen, 2004, A1).

The school system superintendent, as well as many teachers, opposed the new policy as an intrusion of religion into the science curriculum. Critics of the Dover school board's decision contended that the board was trying to insert a Biblical approach to the origin of life into the science curriculum, something they said improperly inserted religion into the public schools. Extended debate ensued. Evolution supporters stated that Darwinian natural selection is universally accepted by mainstream scientists and that intelligent design is a thinly veiled attempt to insert the Genesis version of creationism into the public school science curriculum. Intelligent design supporters contended that intelligent design, while based on the idea that an "intelligent being" created life, never says who that being was, and that intelligent design stops short of inserting religion in the public schools. When it became known that some parents were pursuing a lawsuit over the inclusion of intelligent design in the schools, some of those parents' children were subjected to taunts of "monkey girl" and "ape boy" (Hume, 2007, 183).

The American Civil Liberties Union and Americans United for Separation of Church and State joined in a lawsuit on behalf of 11 parents against the Dover Board of Education because of its policy regarding intelligent design. The *Kitzmiller v. Dover* case gained national attention as the highest-profile test of intelligent design in the schools.

The lawsuit contended that intelligent design was a thinly veiled effort to impose religious-based creationism in the schools. One parent,

Joel Lieb, told a reporter that "a small group of people is trying to push a religion on everybody. It is basically a way of teaching creationism. . . . It doesn't belong in a science class, just the same as evolution doesn't belong in a comparative religion class" (Badkhen).

Testimony in the Dover case, tried before U.S. District Judge John E. Jones III without a jury (called a bench trial), lasted more than a month. Witnesses who testified on behalf of the school board contended that intelligent design, while based on the concept of an intelligent being, was not creationism in disguise. However, the Discovery Institute's own document, called the Wedge Document, listed one of its goals as replacing science with "theistic and Christian science" (*Kitzmiller v. Dover*, 737).

One witness on behalf of the school board was Michael J. Behe, a biochemistry professor at Lehigh University and author of the book *Darwin's Black Box* (1996). His testimony gave the intelligent design version of how life developed, telling the court in painstaking detail, complete with slides shown on a projector, that the evidence supporting intelligent design is apparent in all types of life, from blood clotting to bacteria. "The appearance of design in aspects of biology is overwhelming," Behe testified. He testified that he considered intelligent design to be a scientific approach, although under cross examination he acknowledged that his definition of science was so expansive that astrology—predicting human behavior based on date of birth and the stars—would qualify as science ("Defense Gets Its Days in Court," 2005, 6). Also among those testifying on behalf of the school board was Scott A. Minnich, an associate professor of microbiology at the University of Idaho. He compared intelligent design to discovering a watch and knowing it was created by someone and not the creation of evolution (Goldstein, 2005, A14).

The defense's case was hurt when two school board members, William Buckingham and Alan Bonsell, who had been instrumental in persuading a majority of the board to support the new policy questioning evolution, both testified that they believed the Bible's account of creation to be literally true. Bonsell, under questioning by the plaintiff's lawyers, also acknowledged that he had arranged the purchase of the 50 copies of *Of Pandas and People* for use in the school library and had obtained the funding through money collected at a church and funneled to him through his father (a member of that church), making it apparent that the board's questioning of evolution was really an attempt to teach religion in the classroom (Goldstein).

A witness for the plaintiffs, Brown University biology professor and biology textbook author Kenneth Miller, testified that the Dover policy was "terribly dangerous" because it led students to think that evolution and intelligent design are equivalent approaches to the origin of life (Sunlon, 2005, A1). Some of the parents who sued the school board also testified that they believed the board was trying to impose a religious viewpoint in science class in favor of creationism.

In closing arguments, the lawyer for the school board, Patrick Gillen, called intelligent design the next scientific "paradigm shift," referring to Thomas Kuhn's book (1962) detailing how scientific advancements are made. Kuhn wrote that new ideas are initially dismissed by the scientists who subscribe to the status quo; then a critical shift in balance occurs, with more and more scientists accepting the new idea, until finally the old ideas are discarded and the new ideas almost universally accepted. Kuhn called this process a "paradigm shift." Gillen also told the judge that while some board members believed in the biblical version of creation, the board's actual policy did not attempt to insert religion in the classroom.

Speaking for the plaintiffs, attorney Eric Rothschild ridiculed the contention that intelligent design was not creationism in disguise. "Its essential religious nature does not change whether it is called 'creation science' or 'intelligent design' or 'sudden emergence theory,'" Rothschild said (Goldstein).

In his ruling on the case, Judge Jones drew a parallel between the Dover case and one of the highest-profile cases involving science and the schools, the Scopes "monkey" trial. In that 1925 case, Dayton, Tennessee, teacher John Scopes was tried for violating a state law prohibiting the teaching of the theory of evolution in the public schools. At the heart of the Scopes case was a 1925 Tennessee legislative act prohibiting the teaching of evolution in any publicly funded school. A group of local business leaders, in an effort to draw attention to their sleepy town, and local civil liberties leaders conspired to test the constitutionality of the law. They enlisted Scopes, a science teacher at the local high school, to serve as a defendant.

As predicted, the trial drew national attention, not only for its core theme—science pitted against the Bible—but also because of the larger-than-life lawyers representing each side. Scopes (and opponents of the law) were represented by famed Chicago lawyer Clarence Darrow. The prosecution was handled by former presidential candidate William Jennings Bryan. The highlight of the trial came on the seventh day of the eight-day trial, when Darrow called Bryan as a witness to testify about the inerrancy of the Bible, an exchange that was later dramatized in the Broadway play and movie "Inherit the Wind."[1] Scopes was convicted, although his conviction was overturned on appeal because of a technicality—the judge, rather than the jury as required by Tennessee law, had levied the $100 fine on Scopes.

PREVIOUS COURT CASES

In his Dover decision, Jones also referred extensively to a 1968 U.S. Supreme Court ruling in an Arkansas case, *Epperson v. Arkansas*, in which the court overturned a 1925 state law that made it a misdemeanor to teach evolution in a state-supported school. In the wake of the Scopes case, the Arkansas Legislature passed a law prohibiting schools from teaching that humans had ascended or descended from a lower order of animals. Despite the law, teachers in Little Rock began using a biology textbook that included a chapter on evolution. The state trial court that heard the case ruled that the law was unconstitutional, but the Arkansas Supreme Court ruled in a brief unsigned decision that the law was a legitimate exercise of state power. On appeal, the U.S. Supreme Court said that the law violated the First Amendment's prohibition against the government establishing a religion: "The overriding fact is that Arkansas' law selects from the body of knowledge a particular segment which it proscribes for the sole reason that it is deemed to conflict with a particular religious doctrine; that is, with a particular interpretation of the Book of Genesis by a particular religious group. . . . Government in our democracy, state and national, must be neutral in matters of religious theory, doctrine and practice" (103–104).

Then, in 1987, the U.S. Supreme Court struck down on the same grounds a Louisiana law that required public schools to provide "balanced treatment" of both evolution and the creationist approach from the Bible; the court found that the law was "a sham" that "seeks to employ the symbolic and financial support of government to achieve a religious purpose" (*Edwards v. Aguillard*, 597). The decision stated, "Families entrust public schools with the education of their children, but condition that trust on the understanding that the classroom will not be purposely used to advance religious views that may conflict with the private beliefs of the student and his or her family" (584).

---

[1]The University of Missouri-Kansas City Law School has an extensive website with documents from the Scopes trial at http://www.law.umkc.edu/faculty/projects/ftrials/scopes/scopes.htm.

Both the *Epperson* and *Edwards* decisions relied on what has become known as the *Lemon* Test, from the 1968 U.S. Supreme Court decision in *Lemon v. Kurtzman*. In that decision, the court came up with a three-part test for determining whether a state statute violates the First Amendment prohibition against establishing a religion in public schools. First, the state law had to have a secular, or non-religious, purpose. Second, the statute could neither advance nor inhibit religion. Third, the statute could not entangle religion and government. If a statute violates any of the three standards, it is unconstitutional.

Judge Jones' decision in the Dover case, ruling in favor of the parents and against the school board, relied heavily on the *Epperson* ruling, the *Edwards* ruling, and the *Lemon* Test, and focused on whether requiring students to learn about intelligent design constituted establishment of religion in the schools. Jones wrote that creationists, after the *Epperson* and *Edwards* decisions, tried a new approach: promoting intelligent design. "We conclude," the judge wrote, "that the religious nature of [intelligent design] would be relatively apparent to an objective observer, adult or child" (718). Judge Jones, in a harshly worded critique of the board's policy and of the witnesses who testified on behalf of the board, also wrote in his decision that even the school board's expert witnesses acknowledged that while intelligent design proponents do not specifically state that the intelligent designer is God, there is no other possible explanation: "To be sure, Darwin's theory of evolution is imperfect. However, the fact that a scientific theory cannot yet render an explanation on every point should not be used as a pretext to thrust an alternative hypothesis grounded in religion into the science classroom or to misrepresent well-established scientific propositions" (764). In particular, Jones was critical of the book *Of Pandas and People*, calling it "badly flawed science" that ignored accepted facts and misrepresented both molecular biology and evolutionary processes (744).

In November 2005, the six school board members who had voted for the intelligent design policy were not reelected. The newly elected board members refused to appeal the judge's ruling, and the matter was quietly dropped. But the Dover case has not been the only dispute over teaching evolution in the public schools.

INTELLIGENT DESIGN IN GEORGIA, KANSAS, AND OHIO

In Cobb County, Georgia, the local school board in 2002 ordered a sticker affixed to 35,000 ninth grade science books that stated: "This textbook contains information about evolution. Evolution is theory and not a fact, regarding the origin of living things. This material should be addressed with an open mind, studied carefully, and critically considered." The school board said it took the action after more than 2,000 parents complained about the teaching of evolution in the classroom.

A group of six parents then sued the school board, contending that the placement of the stickers in the texts was part of an attempt to insert religious-based creation into the classroom. During the trial before U.S. District Judge Clarence Cooper, school board members said that the stickers were the result of an attempt to provide "balance" in the science classroom, although they knew that "balance" might lead to discussion of religion in the class (Torres, 2004, A1).

In his decision, Judge Cooper ruled that the stickers challenging evolution constituted an attempt by the school board to impose religion into the public schools (*Selman v. Cobb County*, 2004). Cooper wrote in his decision, "There is no evidence in this case that the School Board included the statement in the sticker that 'evolution is a theory, not a fact' to promote or advance religion. Indeed, the testimony of the School Board members and the documents in the record all indicate that the School Board relied on counsel to draft language for the sticker that would pass constitutional muster. . . . Still, the informed, reasonable [person] would perceive the School Board to be aligning itself with proponents of religious theories of origin" (1308). The Eleventh Circuit Court of Appeals, however, overturned Cooper's ruling because of problems the appeals court found with the facts Cooper relied on, and ordered a new trial (*Selman v. Cobb County*, 2006). Shortly after, the board settled the case out of court, agreeing not to put any stickers in the textbooks and also agreeing to pay more than $160,000 in court costs accrued by the parents who pursued the lawsuit (Stepp and Torres, 2006).

Meanwhile the Kansas State Board of Education, empowered when conservatives won a majority of the ten seats in 2004, in 2005 voted 6–4, following a series of hearings, to adopt new science standards in the state's schools. The new standards cited controversy over the validity of the theory of evolution and allowed the teaching of intelligent design in science classrooms, stating that fossil evidence supporting the idea of evolution was incomplete and citing "a lack of natural explanations for the genetic code" (Cavanagh, 2005, 1). Although the new standards did not specifically call for the teaching of intelligent design, they did adopt the intelligent design concept that some biochemical systems are "irreducibly complex." Board chairman Steve Abrams framed the debate over evolution as a free speech matter and said that most Kansas residents supported the questioning of evolution:

"Science has always been about free speech and open discussion— except in the area of evolution. Evolution has always been treated as dogma" (Cavanagh, 2005, 1).

Even George W. Bush and Bill Frist, then president and Senate majority leader respectively, weighed in on the Kansas debate, saying students should be taught to question the theory of evolution. Bush told reporters, "I think that part of education is to expose people to different schools of thought," including intelligent design (Dennett, 2005). A 2005 Scripps Howard/Ohio University poll found that half of the 1,005 adults questioned supported Bush's suggestion that public schools should also teach intelligent design, and 54 percent said that they believed God had created the universe and humans in a six-day period.

In 2006, however, Kansas voters removed the state school board members who pushed for the new policy, and the newly elected board voted immediately to revoke the intelligent design language and called for the teaching of evolution as the only scientific approach to the origin of life.

In Ohio, the State Board of Education, at the request of intelligent design supporters, undertook an examination of state science standards in 2002. Among the witnesses before the board were a number of intelligent design proponents, while scientists who supported evolution were not allowed an opportunity to present their side. The debate lasted for weeks, and both sides dug in. The presidents of the 13 public universities in Ohio wrote a joint statement opposing the teaching of intelligent design as an alternative to evolution. In the end the board did not adopt intelligent design as a legitimate scientific approach. It did, however, approve a policy requiring a critical review of evolution and included the statement, "The intent of this indicator does not mandate the teaching or testing of intelligent design." The Discovery Institute praised the Ohio policy as a way to use the "teach the controversy" approach to questioning evolution without imposing a religious viewpoint in public schools. However, in the wake of the Dover, Pennsylvania, court decision and the threat of a similar lawsuit in Ohio, in 2006 the board revoked its earlier policy and reinstated evolution as the proper scientific approach (Rudoren, 2006, A14). The Ohio board's policy was hailed by intelligent design critics, including board member Martha Wise, who said, "It is deeply unfair to the children of this state to mislead them about the nature of science." But the associate director of the Discovery Institute's Center for Science and Culture called the decision "a slap in the face" of Ohio residents, who he said polls showed were overwhelmingly opposed to teaching evolution in the schools (Rudoren).

OTHER COURT CASES

In 2006, Frazier Mountain High School in southern California agreed to stop covering intelligent design in a philosophy class taught by a minister's wife after a hearing was scheduled in a court challenge of the practice. The school's principal had said the purpose of the class was "to help students apply critical thinking to questions about evolution and Intelligent Design" (Hedlund, 2005). But when 11 parents filed a lawsuit challenging the class, the school district canceled the course.

In *Freiler v. Tangipahoa Parish Board of Education* (1999), the Fifth Circuit Court of Appeals found unconstitutional a policy the school board had adopted five years earlier requiring any teacher covering the theory of evolution in either elementary or secondary school to read-aloud the following policy:

> It is hereby recognized by the Tangipahoa Board of Education, that the lesson to be presented, regarding the origin of life and matter, is known as the Scientific Theory of Evolution and should be presented to inform students of the scientific concept and not intended to influence or dissuade the Biblical version of Creation or any other concept. It is further recognized by the Board of Education that it is the basic right and privilege of each student to form his/her own opinion and maintain beliefs taught by parents on this very important matter of the origin of life and matter. Students are urged to exercise critical thinking and gather all information possible and closely examine each alternative toward forming an opinion. (821)

About seven months later, a group of parents filed suit in U.S. District Court challenging the policy on the grounds that it was religiously based, in violation of the First Amendment. The court ruled against the school, partly because board members had discussed teaching creationism during the same meeting at which they had adopted the policy in question. The Fifth Circuit Court of Appeals upheld the District Court decision, finding that "[t]he benefit to religion conferred by the reading of the Tangipahoa disclaimer is more than indirect, remote, or incidental. As such, we conclude that the disclaimer impermissibly advances religion" (348).

In 2009, the Texas Board of Education, after extensive debate, adopted a new policy that would allow teachers and students to debate the validity of evolution; however, it stopped short of requiring them to address the "strengths and weaknesses" of the theory, which is critical language for intelligent design supporters (Guervara, 2009). The Texas board is an important group nationally because of the sheer size of the Texas school system. Textbook manufacturers who want their books chosen for Texas have to publish texts that meet Texas's state-wide standards.

In 2008, the Louisiana Legislature passed the Louisiana Science Education Act, requiring the state board of education to help teachers, principals, and other school administrators to create an atmosphere in which students can have "open and objective discussion of scientific theories," a policy widely seen as tacitly accepting intelligent design as a legitimate scientific topic. But in 2009, the Louisiana Board of Education adopted a rule prohibiting public school science teachers from promoting any religious doctrine and requiring any information they present or material they use to be scientifically sound and sup-ported by empirical evidence. But the board did not specifically ban teaching of creationism or intelligent design, and it allowed teachers and local school boards to choose supplemental teaching materials (Deslatte, 2009).

A court in Minnesota ruled against a high school science teacher who took it upon himself to question evolution. Rodney LeVake was hired by Minnesota's Independent School District No. 656. Despite an official school policy requiring him to teach evolution, a policy he said he could not abide by because he was convinced that evolution could not explain the complexity of life. When asked by school officials to write a position paper on how he planned to teach evolution, LeVake wrote:

> I don't believe an unquestioning faith in the theory of evolution is foun-dational to the goals I have stated in teaching my students about them-selves, their responsibilities, and gaining a sense of awe for what they see around them. I will teach, should the department decide that it is appropriate, the theory of evolution. I will also accompany that treat-ment of evolution with an honest look at the difficulties and inconsisten-cies of the theory without turning my class into a religious one.

The school then removed LeVake from teaching tenth grade biology and reassigned him to ninth grade natural science. LeVake sued the school system for violating his freedom of speech, freedom of religion, and academic freedom. He lost in Minnesota District Court, and the Minnesota Court of Appeals also ruled against him in 2001 (*LeVake v. Independent School District No. 656*, 2001, 506).

The evolution and intelligent design debate has not been limited to secondary schools.

In an incident in Oklahoma case, an adjunct professor at Oklahoma City Community College was not rehired for the 2010–2011 school year after a student complained he "glossed over" evolution in his biology classes and taught creationism and intelligent design instead. The teacher, Michael Talkington, denied teaching creationism or intelligent design but said he "simply acknowledged that there are other schools of thought. I did not teach creationism. I did not promote one view over another" (Pemberton, 2010, 11A). However, intelligent design and evolution are taught at dozens of public universities, not in science classes but in courses such as philosophy and honors classes, where both sides are discussed and their strong points and weaknesses are examined.

The U.S. Senate even waded into the debate over teaching intelligent design in the schools when Rick Santorum, at that time a senator from Pennsylvania, introduced a proposed amendment to the federal No Child Left Behind Act that was designed to impose minimum standards in the nation's schools. Santorum's amendment, which was approved by the Senate but never adopted by the House, and therefore killed, would have required students to be taught to understand alternatives to evolution theory ("Pennsylvania Polka," 2006).

ARGUMENTS FOR AND AGAINST TEACHING INTELLIGENT DESIGN

Regardless of the repeated setbacks in the courts and the decisions by numerous public school administrators to step back from teaching intelligent design as an alternative to the theory of evolution, several authors have continued to argue that intelligent design belongs in the schools. Francis J. Beckwith, for instance, argues in his book *Law, Darwinism, and Public Education: The Establishment Clause and the Challenge of Intelligent Design* (2003) that intelligent design is not creationism because it does not specifically state that God is the designer of life, and therefore the teaching of intelligent design in the schools does not violate the First Amendment prohibition against establishing a religion in the public schools. Likewise, David K. DeWolf, John G. West, and Casey Luskin, writing in the *Montana Law Review* (2007), state, "It is important from the outset to understand that labeling ID 'creationism' simply because many of its proponents believe God created the universe would define the term so broadly as to make it largely meaningless" (19). They argue that rather that being based on the Biblical idea that God created life as described in the book of Genesis, intelligent design is "a recognition that the mechanisms proposed by neo-Darwinism could not adequately explain the informational and irreducible properties of living systems that were increasingly being identified in biological literature as identical to features common in language and engineered machine" (20). Because of that difference, they contend, intelligent design is not imposing a religious viewpoint in the schools but merely is an alternative to Darwinian evolution. Luskin, writing in 2005 on the web site Beliefnet.com, cited the "urban legend" that intelligent design is really creationism because life is so complex only God could have created it. He wrote that intelligent design "is a bona fide scientific theory, and there is nothing unconstitutional about teaching about intelligent design in the science classroom" ("It's Constitutional but Not Smart to Teach Intelligent Design in Schools," 2005).

Further, intelligent design proponents such as the Discovery Institute have taken a new approach to getting intelligent design into the schools. The newest approach is to contend that the concept of academic freedom means teachers should be free to bring intelligent design into the classroom. Academic freedom is the idea that teachers should be free to research and discuss controversial issues without interference from administrators or outside political groups. For the Discovery Institute, refusal to allow teachers to bring intelligent design into the classroom is a violation of academic freedom, which is "too important to be sacrificed to the intolerant demands of extremists on any issue" ("Join the Free Speech Campaign"). Beckwith (2003), for example, argues that it is within a teacher's rights to academic freedom to supplement instruction about evolution with material about intelligent design. Indeed, the U.S. Supreme Court in *Keyishian v. Board of Regents* in 1967 protected academic freedom of college instructors, stating, "Our Nation is deeply committed to safeguarding academic freedom, which is of transcendent value to all of us, and not merely to the teachers concerned. That freedom is therefore a special concern of the First Amendment, which does not tolerate laws that cast a pall of orthodoxy over the classroom" (603). But Jay Wexler (2006), who has written extensively about the law and education, writes that the Supreme Court has never extended academic freedom to secondary schools, contending that while teachers do have a First Amendment right to speak on intelligent design outside of school, there is no corresponding right to speak in the classroom on matters that are not part of the curriculum. Regarding the idea that teachers could use academic freedom to teach intelligent design in science class, he writes, "Teachers could teach their personal views on a whole smorgasbord of controversial topics, and the school board would be unable to stop them" (102).

Another approach is that intelligent design should be included in the science curriculum because intelligent design is controversial, and science classes should include discussions of what is controversial as part of a liberal education (Nord, 2003). DeWolf, writing in the *University of St. Thomas Journal of Law & Public Policy* (2009), notes that unlike the theory of gravity, the theory of evolution is controversial because some people criticize it; therefore, because of that criticism, evolution is controversial and that controversy should be taught in public schools: "To assert that there is controversy over Darwinism is simply to state the obvious. Darwin's theory is controverted scientifically, and because of its implications, it remains controversial for purposes of public education. One rarely sees a headline reaffirming the heliocentric arrangement of the solar system or the theory of gravity. However, one regularly sees headlines announcing that new transitional fossils have been found" (327). Critics of this "teach the controversy" approach to questioning evolution say that its proponents' argument is self-fulfilling; the only people who question the theory of evolution are those who want to question the theory of evolution so that intelligent design will be addressed in the classroom. Evolution may be controversial in social or cultural settings, but for scientists, there is no controversy because evolution is accepted as the only legitimate answer to how life has developed (Petto and Godfrey, 2007). Anthropologist and Director of the National Center for Science Education Eugenie C. Scott also calls the "teach the controversy" movement misleading: ". . . [I]ts proponents mean 'teach that scientists dispute evolution,' which is clearly not the case. Students should not be misled into thinking that there is some great turning-away from evolutionary science by scientists" (2007, 91).

Opponents of teaching intelligent design in the schools also equate the concept with religiously based creationism and say that teaching it in the schools devalues the scientific approach: "Christian conservatives thus have been legally thwarted when attempting to inject intelligent design into science curricula to all informed citizens" (Sharpes and Peramas, 2006, 158). In the edited book *Not in Our Classrooms: Why Intelligent Design Is Wrong for Our Schools*, Scott writes that intelligent design is simply creationism in new clothes. Further, the book argues, "teaching the controversy"—the idea that evolution is not settled science and that holes in the theory cast doubt on it—is an attempt to bring religion into the classroom: "Antievolutionists are realizing that teaching creation science or intelligent design may be superfluous if teaching straight antievolutionism will do the job of discrediting evolution in favor of a religious view" (Scott, 2006, 27).

Opponents of teaching intelligent design as an alternative to the theory of evolution also argue that to be a scientific approach, intelligent design would have to be testable and subject to being found to be false. Intelligent design is not verifiable; therefore it is not science. "What schools should do is teach evolution emphasizing both its successes and still unexplained limitations," writes James Q. Wilson in his chapter "Faith in Theory: Why Intelligent Design Simply Isn't Science" in *Intelligent Design: Science or Religion.* "Evolution, like any other scientific theory, has problems," he notes. "But they are not the kind of problems that can be solved by assuming an intelligent designer (whom ID proponents will tell you privately is God) created life" (51).

CONCLUSION

Whether intelligent design is creationism in disguise or a legitimate scientific alternative to the theory of evolution, there is no doubt that some intelligent design proponents will continue in their efforts to persuade state and local school boards to require public school science teachers to question evolution in the classroom, either as an expression of academic freedom or as an attempt to "teach the controversy." As the leading pro-intelligent design Discovery Institute Center for Science and Culture puts it on its web site:

> Although pro-ID organizations do not advocate requiring the teaching of intelligent design in public schools, they also believe there is nothing unconstitutional about voluntarily discussing the scientific theory of design in the classroom. Pro-ID organizations oppose efforts to persecute individual teachers who may wish to discuss the scientific debate over design in an objective and pedagogically appropriate manner.

Vicki Johnson, writing in *The Educational Forum*, a publication aimed at teachers, also forecasts more efforts to question evolution in the school: "The preponderance of evidence suggests that the battles between intelligent design proponents and the defenders of evolution will continue unabated—a prime example of an enduring cultural conflict in American education" (2006, 234). All told, it seems clear that efforts to question evolution in the schools will continue.

CHAPTER 4

How the Media Depict Intelligent Design

Mass media "framing" of issues, or the ways in which journalists depict issues, has been a growing area of research by communication scholars. It is a simple idea, when boiled down to its basics: almost everything covered by the mass media is so complex that journalists must reduce the issues to commonly known and easily accessible frames. For instance, during the economic downturn during the final days of President George W. Bush's term in office and the early days of President Barack Obama's term, did the federal government spend billions of dollars on a bailout of Wall Street firms, or did it spend billions of dollars on a rescue of Wall Street firms? Each of these was media framing at its most basic: Bailout, or rescue? Which frame is used casts the debate in different terms. In other words, the media do not serve as a mirror reflecting all of reality. Rather, the media serve as a window through which media consumers see only a small segment of reality. That window frame through which media consumers see the world provides a rich venue for research.

WHAT FRAMING IS

Robert Entman (1993) writes that frames focus on and highlight some aspects of the subject of communication, raising their importance. Framing involves selecting an aspect of an event and making it more salient "in such a way to promote a particular problem definition, causal interpretation, moral evaluation, and/or treatment recommendation" (52). Facts alone have no meaning of their own. It is only through being placed in some context through emphasis or focus as part of a frame do facts take on relevance (Gamson, 1989). To William Gamson and Andre Modigliani (1989), a frame is a central organizing idea or story line that provides meaning to reports about an issue. "The frame," W. Russell Neuman, Marion Just, and Ann Crigler (1992) write, "does not predetermine the information individuals will seek but it may shape aspects of the world that the individual experiences either directly or through the news media and is thus central to the process of constructed meaning" (61). Frames are not selected by the media accidentally, but instead actively selected as a way to allow consumers to interpret and discuss events (Tuchman, 1978).

Framing is not monolithic. A single news story may contain several frames, as reporters, editors, and sources attempt to highlight aspects of events. By selecting whom to quote and what information those quotes convey, a journalist can actively frame a story (Tuchman, 1972). Gamson (1989) writes that there may be many "senders" in most news reports: "The reporter or anchor person suggests a story line in the lead and closing; the sources quoted suggest frames in sound bites or interviews. . . . For many events there may be more than one frame suggested, and one needs to ask questions about the prominence of competing frames in the same news report" (158). A single sentence may contain several frames, or none at all (Entman, 1993). Frames are often unspoken and unacknowledged (Gitlin, 1980). James Tankard, Jr., (2001) states that the audience is often unaware that framing is taking place; he writes, "Media framing can be likened to the magician's sleight of hand—attention is directed to one point so that people do not notice the manipulation that is going on at another point" (97). The selection of frames has a crucial impact on the way consumers of media reports see the world. As Zhongdang Pan and Gerald Kosicki (1993) write, "Choices of words and their organization into news stories are not trivial matters. They hold great power in setting the context for debate, defining issues under consideration, summoning a variety of mental representations, and providing the basic tools to discuss the issues at hand" (70).

Several studies have found a connection between media frames and media consumer attention to or memory of information, although the relationship is not exact; individuals have their own cognitive framework in which they receive and interpret media messages (Graber, 1988; Neuman, Just, and Crigler, 1992; Sotirovic, 2000). "On the one hand," Gamson writes, "we have a system of media discourse that frames events and presents information always in some context of meaning. On the other hand, we have a public of interacting individuals who approach media discourse in an active way, using it to construct their own personal meanings about public events and issues" (Gamson, 1988, 165). Pan and Kosicki (1993) state that framing analysis "views news texts as consisting of organized symbolic devices that will interact with individual agents' memory for meaning construction" (58). Still, news media framing can influence public opinion and therefore public policy (Andsager, 2000).

One popular perception of framing is that it is no more than media "spin," or an attempt to depict an issue in a way favorable to one position. At presidential candidate debates, the media room, where print and broadcast—and increasingly, Internet—reporters detail the event is swarmed by political operatives attempting to "spin" stories to show that their candidate won the debate. But framing is more than spin. Spin involves knowingly and purposely attempting to sway public opinion; framing involves conscious or unconscious casting of a news story to make the event understandable in the context of the larger picture.

FRAMING OF INTELLIGENT DESIGN AND EVOLUTION

Perhaps no area of media framing is more suited to examination than coverage of the intelligent design-evolution debate. Proponents of the theory of evolution state that the scientific method and advances in the understanding of evidence show that the evolution of life forms is widely accepted based on the discovery of fossils and the wide variety of organisms. Proponents of intelligent design, for example William Dembski, use Mount Rushmore in South Dakota as an exemplar of the concept; as Dembski writes, "What about this rock formation convinces us that it was due to a designing intelligence and not merely wind an erosion?" (2004, 33). Intelligent design proponents argue that life is too complex to have been created by natural—non-intelligent—means.

Framing of the intelligent design-evolution debate is important because news media accounts may be the only exposure most people have to the issue, and it is only through the frames that the media choose that the public can make decisions about what to believe. As Matthew Nisbet and Chris Mooney (2007) put it, many scientists believe that if they simply explain evolution better, their position will be accepted as truth: "In reality, citizens do not use the news media as scientists assume. Research shows that people are rarely well enough informed or motivated to weigh competing ideas and arguments" (56). They bring to the table their own religious beliefs and information from web sites and other sources that match their own outlooks.

Nisbet (2010) calls media framing of science a new paradigm in public engagement, in particular approaching the issue from the perspective of how those who create the news—as opposed to reporting it— frame the intelligent design-evolution debate. Nisbet writes about a gathering of scientists in Washington, D.C., in 2008 to decide how they would frame the issue to gain the widest public support. Rather than framing the dispute in terms of court precedents and the First Amendment prohibition on establishing an official government religion, they found the most effective frame was to cast the debate in terms of how evolutionary science is essential for advances in medicine. In this way, the policy makers help to set the frames that the news media use in reporting about the issue—what mass communication researchers refer to as "setting the agenda."

Media scholars nearly a century ago relied on the "hypodermic needle" or direct effects model of media influence, in which media messages led directly to behavioral changes; this kind of misguided idea of how the media influence society led some to want to restrict comic books because of the belief that teens reading about crime in a comic would invariably turn into drug-using juvenile delinquents. Modern media research, however, has led to a much more nuanced idea of media effects. In agenda setting, the media can not tell people what to think, but can only lead them to the topics about which they will form their opinions. By framing support for evolution as support for medical advances, the scientists cited by Nesbit are attempting to set the agenda in a way favorable to the scientists' position.

Nisbet cites the intelligent design-evolution debate as a classic example of framing an issue by all sides. Even the term "intelligent design" is an attempt by evolution opponents to frame their opposition in a more palatable form after attempts to require public schools to teach creationism were ruled unconstitutional by the courts. He writes that when the news media began to pay attention to the intelligent design movement, the media adopted several frames that were advanced by intelligent design proponents. Nisbet and Chris Mooney (2007) depict several of these frames. First, intelligent design proponents cast their approach as an alternative to scientific uncertainty about evolution. Next, they frame the issue in terms of public schools merely teaching about the controversy they had created over evolution. Scientists who support the theory of evolution, on the other hand, attempted to frame the dispute in terms of the technical aspects of evolution.

Nisbet (2010) writes that the news media ignored or deemphasized the scientists' frames, writing news stories and opinion pieces that focused on frames of strategy and conflict; in other words, they adopted as legitimate frames approaches that cast the intelligent design proponents as dealing in "scientific uncertainty" and "conflict." In doing so, the news media "would carefully balance arguments from both sides, thereby lending credence to the claim by ID proponents that there was growing 'scientific controversy' over evolution, when in fact, there was none" (52).

Mooney and Nisbet (2005), writing in *Columbia Journalism Review*, called television news coverage of the intelligent design movement particularly troubling because of television's need to simplify issues and its reliance on visual images make coverage susceptible to easy to- convey and -understand frames: "Even the best TV news reporters may be hard-pressed to cover evolution thoroughly and accurately on a medium that relies so heavily upon images, sound bites, drama, and conflict to keep audiences locked in. These are serious obstacles to conveying scientific complexity. And with its heavy emphasis on talk and debate, cable news is even worse.

The adversarial format of most cable news talk shows inherently favors ID's attacks on evolution by making false journalistic 'balance' nearly inescapable" (34). Their study specifically examined letters to the editors, opinion columns, and editorials in newspapers both in communities where there was already a public fight over teaching evolution in the schools and in larger publications such as *The New York Times*. They found that intelligent design advocates often "hijacked" the newspaper opinion sections with a flood of pro-intelligent design and antievolution letters. "Without a doubt, then, political reporting, television news, and opinion pages are all generally fanning the flames of a 'controversy' over evolution. Not surprisingly, in light of this coverage, we simultaneously find that the public is deeply confused about evolution" (38), they write, citing public opinion polling demonstrating confusion over whether evolution is accepted science.

Another examination of how the news media depict the intelligent design-evolution debate was published by scientists Jay Rosenhouse and Glenn Branch in *Bioscience* (2006). Their research looked exclusively at news coverage of the attempts to teach intelligent design in the classroom. They found that little science news makes it into newspapers, with most stories on the dispute assigned to education or political writers who have no background or expertise in science. They also write that journalists' need for conciseness distorts the positions of the two sides, making it appear that intelligent design is not religion-based but is a science-based critique of evolution. They also write that newspaper coverage frames the intelligent design movement as an outgrowth of cultural conservatism and religious fundamentalism. Rosenhouse and Branch also examined news magazine coverage, writing that much reporting in magazines such as *Time* and *Newsweek* is framed in a way that both sides have "dueling quotes," with both sides and "a regrettable tendency to frame the debate in preferred terms of the ID side" (249).

Rosenhouse and Branch's study is a rarity in that it also addresses television news coverage of intelligent design and evolution (primarily cable news coverage); they found even more simplification of the issues than in print reporting because of the time constraints of television. On cable talk shows such as *Hardball* on MSNBC or *The O'Reilly Factor* on Fox, the authors write, debate over the theory of evolution and intelligent design is particularly superficial, and the debate is often framed in religious terms with evolution supporters cast as being anti-God while intelligent design supporters are cast as merely asking that evolution should be questioned.

Justin Martin et al. (2006) also examined news media framing of the intelligent design movement, examining 575 newspaper articles from major U.S. newspapers from 1986 to 2003. Their study found that a majority of news articles, newspaper editorials, opinion columns, and letters to the editor about intelligent design cast intelligent design as a religious movement (38%) or a combination of religion and science (37%) as opposed to a purely scientific movement. In addition, a majority of the articles framed intelligent design as a creationist approach (58%), while a minority referred to intelligent design as a fundamentalist approach (12%).

Joshua Grimm, writing in *Science Communication* (2010), studied newspapers in cities in which the intelligent design debate involved local schools: in Kansas, Ohio, and Pennsylvania. The main frames he found were that evolution is controversial (19%), intelligent design is creationism in disguise (12%), evolution is not controversial (9%), and neutral or procedural of how the decision would be made (46%). Grimm reported that intelligent design sources in the articles he examined primarily used one consistent frame: evolution is controversial and should be questioned in the schools. Evolution supporters used two frames: evolution is not controversial, and intelligent design is religion in disguise.

In a study published in *Journalism and Mass Communication Quarterly* (2010), Edward Caudill looked at 235 newspaper articles, editorials, and opinion columns published during 2005, when the debate over the *Kitzmiller v. Dover* case in Pennsylvania and the Kansas State Board of Education dispute over intelligent design were being discussed in newspapers across the country. Caudill found that the news articles and opinion pieces treated the Dover and Kansas cases as political issues pitting conservative Christians against those people opposed to teaching intelligent design in the public schools. He also found that the press made an effort to treat intelligent design arguments and support for evolution even-handedly. Caudill also examined television news coverage of Dover and Kansas, reporting that newscasts also treated the dispute as a political issue, and he reported that news magazines such as *Time* and *Newsweek* framed the disputes as between religion and science, with one *Newsweek* article framing the Dover case as a repeat of the Scopes "monkey" trial. In addition, Caudill noted the intelligent design-evolution debate was often framed as part of a "culture war" between conservative religious groups and scientists, who were framed as being more interested in theory than in God.

Several other researchers who have examined media framing of evolution and intelligent design have been graduate students who examined media coverage as part of their master's degree theses. One of these was Chance York at the University of Kansas (2010). York studied 421 newspaper articles in twelve U.S. cities from 2000 to 2009. He found that 50 percent of the articles he examined framed intelligent design as a religious/unscientific concept and 21 percent framed it as a scientific but not religious approach, while 28 percent were balanced or neutral on the question of whether intelligent design is primarily religious or scientific. But in further analyzing his results, York found that the dominant media frame shifted over time. In the earlier newspapers accounts that York examined, most articles framed intelligent design as primarily a scientific approach, but as time went on, more articles framed intelligent design as a religious issue. He attributed the change to media attention paid to the *Kitzmiller v. Dover* trial in Pennsylvania in 2005.

Another thesis addressing media framing was Justin Martin's in 2004, examining news coverage from 2000 to 2004 of the dispute over whether to include intelligent design in the Ohio public school system. Martin examined 266 news articles, letters to the editor, opinion columns and editorials concerning the Ohio dispute in major U.S. newspapers. Martin's study found that nearly half of the articles about the Ohio dispute were neutral, while 39 percent were negative and just 12 percent positive. He chose as the topic he would examine were whether the newspapers would frame the intelligent design question before the Ohio board of education as more science, religion,

educational consequences, political involvement or social consequences. Martin found that the predominant frame cast the debate as about
science (in 82% of the articles), followed by frames about education
consequences (in 60% of the articles), religion (51%), political involvement (17%), and legal consequences (5%). He also found that writers
used the terms creationist in half the articles examined, despite the
insistence by intelligent design proponents that intelligent design is
not creationism in disguise.

Another study, while not specifically about intelligent design, examined how Tennessee newspapers framed the 1996 state legislature's
debate over whether evolution should be taught in the state's public
schools. The study (McCune, 2003) found, not surprising because
Tennessee was home to the infamous Scopes trial, that the dominant
newspapers frame cast the legislative debate in terms of the Scopes
trial. Other major frames included allegations that the bill was the
result of right-wing politics, that opponents questioned its constitutionality, and that critics said it was an attempt to intimidate teachers.
Supporters of the bill framed the debate as an issue of local control of
the schools, a matter of morals and protecting children.

A FRAMING STUDY EXAMINED

Another study of newspaper framing of the intelligent design-
evolution debate examined coverage of the issue in major U.S. newspa-
pers, including *The Atlanta Journal-Constitution, The Baltimore Sun, The
Boston Globe, The Chicago Sun-Times, The Christian Science Monitor, The
Cleveland Plain-Dealer, The Columbus Dispatch, The Hartford Courant,
The Los Angeles Times, The Milwaukee Journal-Sentinel, The Minneapolis
Star-Tribune, The New York Daily News, The New York Times, Newsday,
The Pittsburgh Post-Gazette, The Rocky Mountain News, The Sacramento
Bee, The San Diego Tribune, The San Francisco Chronicle, The Seattle
Times, The St. Louis Post-Dispatch, The St. Petersburg Times, USA Today,*
and *The Washington Post.*

Five major frames dominated news coverage of the intelligent
design-evolution debate. More than half the articles framed the movement to have intelligent design in the classroom as a covert effort to
teach creationism in public schools. Almost half framed the debate in
terms of evolution as a scientifically proven and a fundamental principle of science. About a third framed the debate over evolution as a
move to teach religion in the public schools, and a third framed it as a
revisiting of the Scopes monkey trial. A quarter of the articles framed

the debate in political terms, linking the intelligent design movement to cultural conservatism.

Intelligent Design Is Really Creationism

Fifty five percent of the articles framed proponents of teaching an alternative to the theory of evolution as creationists in disguise. *The Cleveland Plain Dealer*, for instance, led a story on the intelligent design debate (Stephens, 2004) by stating that a tenth grade biology curriculum approved the night before by the state school board prompted fears by scientists that "it will allow creationism into high school science classes" (A1). Three days later, a *Plain Dealer* columnist (Fulwood, 2004) flatly stated that the Ohio state school board "opened the schoolhouse door to proponents of 'intelligent design'—creationists, in other words—by encouraging criticism of evolution in tenth-grade biology classes" (B1). *The San Antonio Express-News* (2009), in a report on state science teaching standards (Scharrer), stated in a front-page news article that the term intelligent design "has become a lightning rod in the debate over the teaching of evolution with long-time critics arguing scientific evolution theory is well tested and attempts to dilute evolution are little more than a back-door way to elevate creationism or intelligent design" (A1). *The Sacramento Bee* (Rosen, 2004) wrote in a news story that proponents of intelligent design were using a video series to support their argument, and the videos were sold by an organization that promoted both intelligent design and creationism. *The St. Louis Post-Dispatch* published an op-ed piece (McClellan, 2004) stating that it is "anti-science" to "take evolution out of the schools and teach creationism. If you want to be politically correct, you can call it 'intelligent design'" (C1). *The Minneapolis Star Tribune* (Levy, 2004) stated that a local school board unanimously passed a resolution allowing alternative theories of origins to be taught in science classes but noted, "the words 'evolution' and 'creationism' were never mentioned" (B1). *The Atlanta Journal-Constitution* ("No Matter," 2004) argued in an editorial:

> Years ago, after creationists lost their crusade to ban the teaching of evolution in the nation's classrooms, they decided to shift tactics, gradually repositioning their religious beliefs about the origins of life by dressing it up in scrubs, a stethoscope and other scientific garb and renaming it "intelligent design." But the essence of their argument, that human life is the work of a cosmic designer rather than an evolutionary process of natural selection, remained unchanged. (A18)

USA Today (Parker, 2004) called intelligent design itself an evolution of creationism. *The San Francisco Chronicle* (Badkhen, 2004) wrote that the intelligent design movement in Dover, Pennsylvania, was supported by the Institute for Creation Research, "the world leader in creation science" (A1). *The Washington Post* ("God and Darwin," 2005) wrote in an editorial that intelligent design is a more sophisticated version of creationism. Perhaps most typically, *The Christian Science Monitor* (Dotinga, 2004) simply linked "proponents of teaching biblical creation and intelligent design" (11), as if the two are inextricably linked and there is no further need for elaboration. In a particularly harsh critique of the intelligent design movement, *Tampa Tribune* columnist Daniel Ruth (2008) wrote that intelligent design proponents have as a goal "eventually introducing into public schools the wizard in the sky mumbo jumbo argument for creationism" (Metro 1).

Evolution Is Accepted Scientific Fact

Forty four percent of the articles framed the intelligent design debate in terms of evolution being widely accepted scientific fact, the basis for modern biological science, or a fundamental truth of the origin of life. *The Dallas Morning News* reported (Stutz, 2008) that 95 percent of public and private college science teachers in Texas opposed a proposal to require public secondary schools to teach weaknesses in the theory of evolution. *The Washington Post* (Strauss, 2004) published a news article stating, "the vast majority of scientists agree that evolution is a proven unifying concept in science" (A14). *The Chicago Sun-Times* (Begley, 2004) published an entire article on a biology teacher at a "fundamentalist Christian college" (Olivet Nazarene University) who, despite opposition from many at the school, teaches evolution because, he said, it "has stood the test of time and considerable scrutiny" (20). *The Baltimore Sun* (Hirsch, 2005) published a news story stating that "the core elements of evolutionary theory are embraced overwhelmingly by mainstream scientists" (6B). *The Washington Post* (Anderson, 2005) wrote that evolution and natural selection "are part of the foundation of modern science" (B3). *The Cleveland Plain Dealer* (Stephens and Mangels, 2005) stated flatly, "Darwin's widely accepted theory holds that life on Earth descended from common ancestors and changed over time due to natural selection" (B4). *Newsday* (Riley, 2005) wrote that intelligent design has "little support in the mainstream biology community" (A10). *The Los Angeles Times* ("Academic Hall of Shame," 2004) editorialized that there is "overwhelming evidence supporting the widely accepted theory of evolution" (B20). *The Boston Globe* (Mishra,

2004) wrote in a news story that evolution is accepted, and "most scientists consider it to be among the most important and well-supported scientific theories of all time" (A1). *The Atlanta Journal-Constitution* (Torres, 2004, November 7), in an article about the Cobb County, Georgia, school board's decision to attach stickers to biology texts questioning the theory of evolution, stated in the lead—or first sentence—that evolution is "a widely accepted scientific theory" (A1). *The New York Times* (Glanz, 2004) referred to mainstream evolution, *The Columbus Dispatch* ("Debate Keeps Evolving," 2004) editorialized that most scientists reject intelligent design, and *The Baltimore Sun* ("Fear of Evolution," 2005) editorialized, "As far as scientific theories go, evolution is more than sound" (A14). *The New York Times* (2008) editorialized that those who want schools to teach the weaknesses of the theory of evolution are "foolish" and that students "will likelier emerge with a shakier understanding of science" (18). Daniel Ruth's column in *The Tampa Tribune* (2008) compared the theory of evolution to "Newton's theory of gravity or relativity" (Metro 1).

Intelligent Design Is Really Religion in the Schools

Thirty five percent of the articles framed the intelligent design movement as an attempt to insert religion, particularly conservative Christianity, into public schools. For example, *The Sacramento Bee* (Rosen, 2004) began a news article about an attempt to persuade a local school board to remove the theory of evolution from the science curriculum with a vignette of a packed school board meeting: "One man wore a T-shirt that showed a monkey beneath the words: 'Our father?!' Under the monkey was this statement: 'The wisdom of men is foolishness to God' " (N1). Fulwood (2004) wrote in a column in *The Cleveland Plain Dealer*, "Proponents of ID are determined to push their fundamentalist Christian beliefs into schools" (B1). A *Cleveland Plain Dealer* editorial (2006) noted that while Ohio's board of education decided not to adopt a plan requiring public school science classes to discuss problems with the theory of evolution, stating that "even with intelligent design stricken from Ohio's science classes, no similar obstacle exists to teaching the subject, say, in a religion class" (B8). *The New York Times* (Glanz, 2004) began a news article about evolution and intelligent design with a description of a Baptist minister leading the fight against evolution in Ravalli County, Montana, "which nurtures a deep vein of conservative religious sentiment" (1–22). *The Cleveland Plain Dealer* ("Sweating Details," 2004) editorialized that a plan to place stickers challenging evolution in Georgia textbooks "amounts to an unconstitutional breach of the wall between church and state" (H2). *The Atlanta Journal-Constitution* ("Intelligent Redesign," 2004) editorialized that the intelligent design movement challenges "the cherished and hard-won American principle that public schools will not become pulpits and teachers will not become preachers" (A18). In a news story in *The Atlanta Journal-Constitution* the same day (Torres, 2004, November 10), Cobb County, Georgia, school officials were described in the lead as knowing that their textbook stickers questioning the validity of the theory of evolution would result in religion being brought up in the classroom. *The Seattle Times*, in a column (Westneat, 2004), equated intelligent design with religious conservatives, and *The Washington Post* ("God and Darwin," 2005) editorialized that "to teach intelligent design as science in public schools is a clear violation of the principle of separation of church and state" (A14). *The Denver Post* ("Judge Ruled for What's Right," 2006) editorialized that the judge in the landmark *Kitzmiller v. Dover* court case upheld the separation of church and state, writing, "Religious beliefs on the origins of life have their place in the church, synagogue or mosque. But in the public school classroom, science must rule" (E4).

Intelligent Design Is a Replay of the Scopes Monkey Trial

Thirty four percent of the articles framed the intelligent design-evolution debate in terms of the 1925 Scopes Monkey Trial, in which Dayton, Tennessee, teacher John Scopes was tried for violating a state law prohibiting the teaching of evolution in the public schools. Scopes was convicted in a trial that garnered nationwide news coverage, although his conviction was overturned on appeal. Typical of the Scopes frame was a piece in *The Baltimore Sun* ("Fear of Evolution," 2005), which editorialized that the debate over teaching evolution "sounds like echoes of the Scopes monkey trial 80 years ago" (A14). An op-ed columnist in *The San Diego Union Tribune* (Van Deerlin, 2005) wrote, "Why, we wonder, must serious scientists still fear the sort of harassment visited on Galileo four centuries ago? Or, for that matter, 75 years after the celebrated Scopes trial, when the state of Tennessee put a young high school teacher on trial for teaching evolution?" (B7). *The Boston Globe* wrote in a news story (Mishra, 2004) that the debate over evolution had been "considered a relic of the past" (A1) since the Scopes trial and the play loosely based on it, *Inherit the Wind*. *The Atlanta Journal-Constitution* ("Intelligent Redesign," 2004) editorialized that the debate over evolution and intelligent design is "an unfortunate echo of the famed Scopes 'Monkey' Trial of 1925" (A18).

A *Cleveland Plain Dealer* editorial ("Sweating Details," 2004) called the debate over evolution a "modern-day version of the Scopes trial" (H2). An *Atlanta Journal-Constitution* news story (Torres, 2004, November 7) reported that a looming court battle over teaching evolution "could stir comparisons to the 1925 trial in Dayton, Tenn., when John Scopes was tried for teaching evolution" (A1). *USA Today*, in a news story about teaching evolution in public schools (Parker, 2004), began: "The long-simmering battle over how evolution is taught in high school biology is boiling again. Nearly 80 years after the famous 'Monkey Trial,' in which Tennessee teacher John Scopes was convicted of teaching evolution in violation of state law, 24 states this year have seen efforts to change the way evolution is taught" (A3). *The Christian Science Monitor* (Sappenfield and McCauley, 2004) tied the battle against teaching evolution to the Scopes trial, and *The Washington Post* (Strauss, 2004) wrote that "eighty years after John T. Scopes, a high school biology teacher, was charged with illegally teaching the theory of evolution in Tennessee, the social and intellectual values that imbued that trial with such meaning continuing to stir emotions, prompting challenges in school boards and state legislatures, courthouses and schoolrooms" (A14). John Riley (2005) wrote in *Newsday* that no issue "has been more durable than the Darwin vs. God struggle in the public schools, with a pedigree dating to the famous Scopes monkey trial in Tennessee in 1925" (A10).

Intelligent Design Is a Political Issue

Twenty four percent of the articles framed the intelligent design-evolution debate in political terms. *The New Orleans Times-Picayune* (Barrow, 2008) focused on how a bill that would allow local school boards to require science teachers to discuss the weaknesses of the theory of evolution and ideas about cloning and global warming would help Louisiana Republican Governor Bobby Jindal's political ambitions. Commenting on the same bill, *The New York Times* (2008) in an editorial tied the bill to Jindal as "a rising star on the conservative right" who may be thinking of politics more than science (18). *The New York Times* (Banerjee, 2004) tied support for intelligent design to election campaigns that conservative Christians successfully worked for in 2004. Another *New York Times* article (Jacoby, 2005) referred to "a renewed determination by anti-evolution crusaders—buoyed by conservative gains in state and local elections—to force public schools science classes to give equal time to religiously based speculations about the origins of life" (A19), including intelligent design.

The Boston Globe ("Creationists at the Gate," 2005) connected support for intelligent design to supporters "emboldened by the important role social conservatives played in the re-election of George H. Bush" (A14). *Newsday* (Riley, 2005) published a news article linking support for intelligent design "to the same forces that have helped catapult Bush into the White House" (A10). *The Christian Science Monitor* (Sappenfield and McCauley, 2004) placed support for intelligent design on the red side of the red/blue state division. A column in *The St. Petersburg Times* (Blumner, 2004) tied the intelligent design debate to the "the redder parts of this country" (P7). *The Boston Globe* (Mishra, 2004) reported that most challenges to teaching evolution had been in states carried by Bush in 2004. An op-ed column in *The Hartford Courant* (Cohen, 2004) began, "The praise-the-Lord evangelical revolution that reportedly swept President Bush to his second term was over about 16 seconds when the results were in" (A15). One of those results, Cohen wrote, was the intelligent design movement in public schools. *The New York Daily News* printed a column (Louis, 2004) that linked the intelligent design movement to the religious right. *The Chicago Sun-Times* published a column by Tom McNamee (2007) that referred both to then-president George W. Bush's statement that he supported the teaching of intelligent design in science classes and the 2008 presidential campaign, in which three of the nine people running for the Republican nomination for president said they did not believe in evolution.

TELEVISION COVERAGE

Television coverage of the intelligent design-evolution debate, more than print coverage, is more susceptible to the "he said-she said" form, particularly on the cable talk shows such as *The O'Reilly Factor* on Fox or *Hardball* on MSNBC. Constrained by the time limits inherent in television programming, shows such as these cover the intelligent design and evolution arguments by having a proponent from each side spend a few minutes on camera. This "equal time" approach gives the appearance of being even-handed, of meeting the journalistic ideal of objectivity, but at the same time it frames the dispute as being between two sides that have the same amount of evidence supporting them. While intelligent design proponents argue that their field is scientific, nearly all mainstream scientists disagree and contend that evolution is the only acceptable theory for the origin of life. An example of this "balanced" approach to the issue occurred on the MSNBC *Hardball* show on April 21, 2005, when host Chris Matthews hosted a segment on intelligent design and evolution. His guests were the Rev. Terry Fox of Wichita's Immanuel Baptist Church, who was involved in the effort to persuade the Kansas Board of Education to include intelligent design in the science curriculum, and Eugenie C. Scott, the executive director of the pro-evolution National Center For Science Education ("Hardball with Chris Mathews," 2005). Larry King, on his CNN talk show, strived for balance in an episode that aired August 23, 2005, by inviting two intelligent design proponents and two evolution supporters ("Transcripts," 2005).

Often, though, particularly on shows in which the host has a strong ideological stance, a guest is either selected for a particular viewpoint, agreeing with or disputing the host's ideas. Bill O'Reilly's show on Fox, *The O'Reilly Factor*, is not shy about stacking the deck. On October 12, 2009, for instance, O'Reilly interviewed evolution supporter and acknowledged atheist Richard Dawkins, an author ("O'Reilly vs. Atheist Author Richard Dawkins," 2009). During the interview, O'Reilly gave his support to intelligent design: "I believe in creative design. I believe in evolution, but I think it was overseen by a higher power, because as we just stated and you acknowledged, you guys still haven't figured out how it all began." Likewise, O'Reilly had as a single guest on October 22, 2007, actor Ben Stein, who was promoting his pro-intelligent design documentary, "Expelled: No Evidence Allowed." During the interview, O'Reilly clearly agrees with Stein, much in the same way he disagreed with Dawkins.

Occasionally, long-form reporting on PBS's *NOVA* series will address the intelligent design-evolution debate, as it did in 2009. But one of the more extended discussions of the intelligent design-evolution debate occurred on the Comedy Central channel "news" program *The Daily Show* in 2005 (Pigliucci, 2007). In the series called "Evolution, Schmevolution," *The Daily Show* staff interviewed intelligent design proponents, including Dembski from the Discovery Institute, and evolution supporters, as well as commentators reflecting on how the issue has been addressed in the media. Among those on the series was author and media framing researcher Chris Mooney, who addressed the issue of intelligent design and the theory of evolution in the broader context of politics, religion and science. Host Jon Stewart, of course, often treated the issue with humor, and one of *The Daily Show* regular guests, comedian Lewis Black, pointed out science has helped to create numerous cures for diseases, while religion has helped to create "some pretty good things" such as the Sistine Chapel. Black commented, "The problem is, when you try too hard to apply science to religion, both come off looking ridiculous" (191).

THE INTERNET

Perhaps nowhere is the intelligent design-evolution played out in more detail than on the Web. For every pro-intelligent design web site, there is a pro-evolution site as well. But all web sites are not equal. Of course, web sites, whether created by organizations or by bloggers, are not under the same constraints as traditional print or broadcasting sites. While journalists operate on the idea of objectivity and neutrality, web presences do not operate under the same constraints. They are free to take a stance on an issue such as evolution and intelligent design.

Web content providers—particularly bloggers—often not only create their own content, they also usually link to other sites, including traditional print and broadcasting sites. One study examined how bloggers who focus on science use sources of information. The study (Walejko and Ksiazek, 2010) looked at political- and science-oriented blogs on the topics of intelligent design and global warming. Their study of 41 science bloggers from 2004 to 2007 found that the bloggers often linked to other blogs and online news articles from traditional news media, rather than merely creating their own content.

The list of web sites addressing the intelligent design-evolution debate is too large to print in one book. But here are a few, with a brief description of each. Each of these sites addresses the intelligent design-evolution debate from the perspective of either pro-intelligent design supporters or evolution supporters. There are few sites on the Web that are neutral or even-handed in their approach to the issue.

* www.intelligentdesign.org is an alternative address for the Discovery Institute's Center for Science and Culture, whose web address is also www.discovery.org/csc/. The Discovery Institute is the leading organization in the intelligent design movement. Another Discovery Institute web address is http://www.idthefuture.com.
* www.pandasthumb.org is a science-oriented web site that is pro-intelligent design.
* www.intelligentdesignnetwork.org is a pro-intelligent design site by the Intelligent Design Network Inc., a nonprofit organization.
* www.allaboutscience.org is the site of a group allaboutgod.com and is a religiously oriented discussion of science, primarily from the antievolution viewpoint.
* www.beliefnet.com is an advertiser-supported web site that focuses on all things religious, including intelligent design.

* http://ncse.com is the web site for the National Center for Science Education, a pro-evolution site that includes a large section on intelligent design and evolution.
* www.newscientist.com is a pro-evolution site that also discusses other scientific topics.
* www.nationalacademies.org is the web site of the National Academy of Science and has a large pro-evolution section.

CONCLUSION

One of the standards of journalism in the United States is the concept of objectivity. Journalism textbooks and journalism teachers try to instill in students the idea that reporters should be unbiased, that they should present both sides of a controversial issue so their readers can weigh the evidence and reach their own conclusions about which ideas they should accept and which they should reject.Despite this reliance on the concept of objectivity, it is clear that journalists either consciously or unconsciously use frames as way to make sense of complex issues such as the intelligent design-evolution controversy.

While scientists argue that there is no doubt about the theory of evolution and call it the only generally accepted approach to the origins of species, intelligent design proponents argue that evidence points to a designer creating life forms. Journalists, the vast majority of whom have had no scientific training since their basic science classes in college, usually resort to a "he said this, she said that" approach to reporting about intelligent design and evolution. To aid them in this even-handed approach, they resort to framing their stories in ways that do not necessarily help their audience to understand complex issues.

Judith Buddenbaum (1998) writes that journalists writing about issues involving aspects of religion must take care because "[t]he way journalists frame their stories and organize information can miss the point, baffle the readers or lead them to see bias where none was intended. Their words can enlighten, but they can also misinform, foster stereotypes ..." (172). Stewart Hoover (1998) writes that the ultimate goal of journalists should be to provide a middle ground, clarifying issues rather than focusing on their opposite poles and adding to misunderstanding and conflict.

The numerous framing studies listed in this chapter suggest that rather than providing a middle ground for the discussion of intelligent design and evolution, news articles, opinion columns, and editorials often instead further polarized the issue by choosing frames that discount, criticize, ridicule, or minimalize support for intelligent design. Intelligent design is framed as creation in disguise; supporters are framed as trying to create a new Scopes monkey trial, with intelligent design as religion and not science; and opponents of the theory of evolution are framed as opposing mainstream science.

It is clear from the studies in this chapter that while public opinion polls continue to show that a large part of the American public doubts the legitimacy of evolution, most reporters do not share that skepticism, and their news and opinion articles reflect their own beliefs, either consciously or unconsciously. So why does the public tend to side with the critics of evolution? Has it been because the intelligent design supporters have been more effective in their own framing of the issue? These intelligent design advocates, appearing on television and on the Internet and being quoted in newspaper articles, have attempted to frame the issue as a controversy over evolution. Perhaps that framing by intelligent design proponents has resonated more with the public. Meanwhile, it will be interesting to see whether news media framing of the intelligent design-evolution debate sways public opinion.

CHAPTER 5

Conclusion

A reading of this book should make it clear that there is no middle ground in the debate over intelligent design and the theory of evolution. Supporters either argue that evidence points to a "designer" who is responsible for life, or that the only acceptable scientific approach is evolution and intelligent design is nothing but creationism in disguise.

News media coverage of the intelligent design-evolution argument has framed the issues in terms that are familiar to readers and viewers. As recounted in the previous chapter, the media framing of the issue has cast the debate in terms that generally favor the evolution viewpoint. Intelligent design is framed as an attempt to insert religion into the public schools, or to recreate the 1925 Scopes "monkey" trial in Dayton, Tennessee. The media also have framed the debate in ways that cast doubt on whether intelligent design is a scientific approach, as supporters contend.

Framing of debates over evolution is not new to the intelligent design movement. During the Scopes trial, for instance, famed Baltimore journalist H. L. Mencken characterized the trial in Tennessee as a circus (Witham, 2002, 227). In one account, Mencken wrote,

> In brief this is a strictly Christian community, and such is its notion of fairness, justice and due process of law. . . . Its people are simply unable to imagine a man who rejects the literal authority of the Bible. The most they can conjure up, straining until they are red in the face, is a man

who is in error about the meaning of this or that text. Thus one accused of heresy among them is like one accused of boiling his grandmother to make soap in Maryland. . . . The trial, indeed, takes on, for all its legal forms, something of the air of a religious orgy. (Mencken, 1925)

Indeed, framing is not a new concept.

Of course, framing does not indicate a single-minded approach to any news coverage. Instead, some framing has been favorable to the intelligent design movement. And virtually every account of the debate has multiple frames, some favorable to evolution and some favorable to intelligent design.

Much of the intelligent design movement has been focused either on teaching intelligent design in the public school classrooms or, alternatively, persuading local and state school boards to require that public school science teachers tell their students, either directly or by way of stickers in textbooks, that scientists have questions about the theory of evolution. Arguing for the approach of "teaching the controversy," intelligent design proponents are hoping that students who are told they can question the theory of evolution will seek out answers on their own, and that some of those answers will be found in intelligent design.

While the news media accounts of the intelligent design-evolution debate cast doubt on the scientific or religious nature of intelligent design, public opinion polls show that a significant number of Americans have bought into the idea that evolution should be questioned in the public schools. At least a third of the people responding to the Gallup Poll reported that they believed intelligent design is the answer to how life developed on Earth, and about the same percentage believed that scientists have doubts about the theory of evolution, findings that directly support the intelligent design movement's ideas that the theory of evolution is controversial and that students should be taught about the controversy. These data suggest that intelligent design proponents have succeeded in framing their case in a way that resonates with a large part of the American public, perhaps through appearances on television or through their web sites. Other public opinion polls have found similar results. While the media may be framing the debate in one way, it is clear that the intelligent design movement has made gains in the court of public opinion. It is a different story in the actual court system.

Local courts, federal courts, and the U.S. Supreme Court have consistently ruled that efforts to insert religious beliefs into the public schools constitute a violation of the U.S. Constitution's First Amendment, which prohibits the establishment of an official religion. The most well-known of these decisions, 2005's *Kitzmiller v. Dover*, has been characterized as the high water mark of the intelligent decision movement, much in the way that the Battle of Gettysburg in Pennsylvania was the high water mark for the Confederacy in the Civil War; just as the Confederacy never again got as far North as at Gettysburg, the intelligent design movement never won as large a victory as when the Dover, Pennsylvania, school board voted to teach intelligent design in the public school science classes, only to lose in U.S. District Court. It is ironic that Dover and Gettysburg are just 27 miles apart.

While news coverage of the intelligent design movement may have reached its peak during the legal battle and ensuing court decision in Dover, by no means has the issue dropped from the media's radar. Newspapers, television programs, and Internet blogs and other postings still address the issue with regularity, with coverage often prompted by some new legal case, book publication, or statement by proponents. In addition, as the references cited in this manuscript indicate, the intelligent design and evolution debate has proved to be fertile ground for book publishers. As the framing studies examined in the previous chapter indicate, the Dover, Pennsylvania, court decision marked a turning point in the intelligent design movement, as well as in reporting about intelligent design and evolution. With Judge Jones' decision in the Delaware case, the news media appear to have become more likely to question the intelligent design movement through their framing. In this way, reporters may have unconsciously picked frames that support evolution and disregard the legitimacy of intelligent design.

Certainly, however, media criticism of the theory of evolution did not begin with the development of intelligent design. Edward Caudill recounts at length the opposition to evolution from religious leaders, creationism advocates, and social critics in his 1989 book *Darwinism in the Press*, which was written before the intelligent design movement even began. There has been no shortage of criticism of the theory of evolution, just as there has been no shortage of scientists arguing on behalf of evolution.

So where will the debate go from here? Intelligent design advocates will continue to push for acceptance of their approach both in the arena of public opinion and in the public school systems. While the Discovery Institute's Center for Science and Culture maintains that it does not favor requiring public schools to teach intelligent design in science classes directly, it does argue in favor of allowing teachers to question evolution and, if they are so inclined, to point students toward the idea of a designer. Intelligent design proponents also will continue their efforts, largely futile up to this point, to persuade the scientific community that intelligent design is a legitimate scientific approach, on a par with the theory of evolution.

Meanwhile, evolution supporters will continue to argue that intelligent design is really a new form of creationism that should not be allowed in science classes and that intelligent design is not a legitimate scientific approach because it does not meet the threshold of being testable, one of the primary ways in which scientific theories are examined.

It also is certain that newspapers, television, and the Internet will continue to report on the issue. How these accounts frame the debate will go a long way toward determining which side will ultimately prevail in the court of public opinion. Will the news media continue to frame the issue in terms of intelligent design as a religion-based creationist approach, or will they frame intelligent design as a science? Will they accept the intelligent design approach and frame intelligent design as a legitimate scientific approach? Will they frame the issue in terms of a controversy over evolution, or will they frame evolution as the only accepted scientific approach to how life has developed?

The idea that journalists, whatever their media platform, are neutral reporters of fact is as outdated notion as the idea that reporters, in choosing what to report, tell their audience what to think. Instead, reporters, in choosing what aspects of a story to report and what parts to emphasize, help to set the agenda for the public; they do not tell the public what to think, but they do help the public decide what to think about. In other words, the news media bring salience, or prominence, to the parts of the news coverage they are focusing on, in how they frame their coverage. More and more media researchers and political scientists have been focusing on framing as way of helping to determine what drives public opinion.

Just as there is no doubt that the debate over intelligent design and evolution will continue and that news coverage of that debate also will continue, there is no doubt that researchers will continue to examine how that media coverage is framed.

APPENDIX A

Names in the News

CHARLES DARWIN (1809–1882)

We must, however, acknowledge, as it seems to me, that man with all his noble qualities ... still bears in his bodily frame the indelible stamp of his lowly origin.

Charles Robert Darwin was born in Shrewsbury, Shropshire, England (about 160 miles northwest of London), on February 12, 1809, into an affluent, well-known family. Charles' father, Robert Waring Darwin (1766–1848), was a physician. His mother, Susannah Wedgwood Darwin (1765–1817)—the heir to £25,000—was the daughter of Josiah Wedgwood, one of the founders of the Wedgwood pottery company and a noted opponent of slavery. Darwin, the fifth of six children, was baptized in Shrewsbury's Anglican church. He had one brother (Erasmus, named after his eccentric grandfather), three older sisters (Marianne, Caroline, and Susanne), and one younger sister (Emily Catherine). He attended Shrewsbury Grammar School but preferred hunting and collecting insects to learning Greek and Latin. In the autumn of 1825, he enrolled at Edinburgh University to study medicine. There, he learned taxidermy from a freed Guyanan slave, John Edmonstone; this was a skill that would prove useful to him later. However, he was sickened by watching surgery and dropped out of college.

In 1827, after being scolded by his father that "you care for nothing but shooting, dogs, and rat-catching, and you will be a disgrace to yourself and your family," Darwin enrolled at Cambridge University and began studying to join the clergy. However, he remained interested in collecting insects, and it was at Cambridge that he began to appreciate the vast diversity of species, an appreciation that springs inevitably from an interest in insects. While at Cambridge, he was mentored by Robert Grant, an admirer of Darwin's grandfather Erasmus and a supporter of Lamarck's theory of inheritance of acquired characteristics. Darwin also learned about natural theology, which sought to understand God by studying God's creation (i.e., nature). To advocates of natural theology, God's goodness was visible in the progression of life from "lower" to "higher" forms, culminating in the special creation of humans. At Cambridge, Darwin read William Paley's *Natural Theology*, which argued that nature's complexity was evidence of design and that design required a designer. Although Darwin studied and enjoyed Paley's book, he later rejected Paley's arguments, "now that the law of natural selection has been discovered." While at Cambridge, Darwin lived in the same dormitory rooms that had housed William Paley seventy years earlier. Today, those rooms—which are now offices—are marked by a statue of Darwin in the adjacent garden.

At Cambridge, Darwin had traditional religious beliefs but a noticeable lack of religious zeal. Although he later claimed that some of his time at Cambridge "was sadly wasted," Darwin made his first scientific presentation there (about *Flustra*, a bryozoan), and in 1831 he earned a bachelor's degree in theology, ranking tenth among 178 non-honors students. However, it was a great era of exploration, and Darwin was drawn to the prospect of travel; he hoped to delay his religious career until he had visited exotic locales.

In the 1830s, the British government commissioned the HMS *Beagle* to conduct expeditions "devoted to the noblest purpose, the acquisition of knowledge." The *Beagle*'s captain was the temperamental 26-year-old Robert FitzRoy, and FitzRoy needed a naturalist for the voyage. He wanted the naturalist to be the Cambridge cleric-botanist John Henslow, a friend of Darwin's. Henslow declined FitzRoy's offer but recommended Darwin for the job. Darwin's father did not want him to go on the voyage, but Josiah Wedgwood (Darwin's' uncle) intervened, convincing Darwin's father to let his son go on the cruise. Darwin later noted that he "was resolved to go at all hazards."

The *Beagle* sailed from Plymouth, England, on December 27, 1831. Darwin's experiences were overwhelming—he hiked through a jungle in Brazil, dug up fossils in Argentina, witnessed an exploding volcano, withstood an earthquake that raised shellfish beds 3' above the shoreline, and studied coral reefs. By the end of his voyage, Darwin had written 1,383 pages of notes about geology and 368 pages about zoology; entered 770 pages in his diary; preserved 1,529 specimens; and labeled 3,907 skins, bones, and other specimens. He had also seen numerous interesting geological formations and new habitats (e.g., the Andes and tropical islands). His voyage lasted fifty-eight months, forty-three of which were spent in South America.

While at sea, Darwin read Volume 1 of Charles Lyell's *Principles of Geology*, which had been given to him by Captain FitzRoy. Lyell documented that ancient Earth had been molded by the same slow, directionless forces that shape Earth today. When the *Beagle* docked at St. Jago, Lyell's claim that land slowly rises and falls was confirmed when Darwin found a band of coral fragments and shells high among the volcanic cliffs. Later, when exploring the Andes, Darwin found fossil trees that had grown on a sandy, shell-littered beach. Lyell showed that Earth is very old and that Earth's history had been characterized by the extinction and appearance of innumerable species. This was the world that Darwin sought to explain.

In September of 1835, the *Beagle* docked for a month-long stay at a group of thirteen volcanic islands called the Galapagos Islands. These islands, which straddle the equator, are about 620 miles west of Ecuador. Darwin later noted that the islands' songbirds—called finches—appeared as if "one species had been taken and modified for different ends." Darwin was told by Nicholas Lawson that local inhabitants could tell the home island of a tortoise just by examining its shell. As Darwin noted later, "By far the most remarkable feature in the natural history of this archipelago [is] that the different islands to a considerable extent are inhabited by a different set of beings." This intrigued Darwin, for it suggested that each island had its own group of organisms. Did each island really have a unique species of tortoise? Did each island have a unique species of finch? Had there been a separate creation event at each island? Darwin was initially skeptical: "I never dreamed that islands 50 or 60 miles apart, and most of them in sight of each other, formed of precisely the same rocks, placed under a quite similar climate, rising to nearly equal height, would have been differently tenanted."

After leaving the Galapagos, the *Beagle* sailed to Tahiti, New Zealand, and Australia, during which time Darwin pondered the questions he had been developing about life's diversity. Soon after Darwin returned to England in October 1836, ornithologist John Gould of the London

Zoological Society examined the birds that Darwin and others had collected on the Galapagos Islands and told Darwin that each island *did*, in fact, house a separate species of finch.

Back in England, Darwin lived on Fitzwilliam Street in Cambridge from 1836 to 1837. Today, his residence is marked by a small plaque.

In September of 1838, Darwin read "for amusement" Thomas Malthus' *An Essay on the Principles of Population*. Malthus argued that populations could grow exponentially (e.g., 2, 4, 8, 16, 32, 64, 128, 256, etc.) but that resources such as food can increase only linearly (e.g., 1, 2, 3, 4, 5, 6, 7, etc.). Malthus' book had a tremendous influence on Darwin: What would be the consequences of a constant struggle for existence (as Malthus proposed) that persisted for millions of years (as Lyell proposed)?

As Darwin pondered this question, he realized that Malthus' struggle throughout the history of Lyell's ancient Earth might explain the great diversity of plants and animals that he had encountered on his travels. As he later wrote in his autobiography,

> Being well prepared to appreciate the struggle for existence . . . it at once struck me that under those circumstances, favourable variations would tend to be preserved and unfavourable ones destroyed. The result of this would be the formation of a new species. Here, then, I had at last got a theory by which to work.

Combining Malthus' idea with what he had seen at the Galapagos (which he called the primary source of all his views) gave Darwin a new insight. He now knew how species evolved. He wrote, "Did [the] Creator make all new [species on oceanic islands], yet [with] forms like [on] neighboring continent? This fact speaks volumes. My theory explains this but no other will." Despite his confidence, however, Darwin was not ready to announce his discovery.

As experts continued to sort through Darwin's enormous collection of specimens from the *Beagle* voyage, Darwin began to think more and more about his radical idea. The finches collected from the Galapagos suggested to Darwin that new species could evolve from a common ancestor. Darwin began to think of humans not as an ultimate and special creation but instead as merely one more species, albeit one with unusual mental powers.

In 1837, Darwin moved to a residence on Great Marlborough Street in London, where he began writing about his "dangerous" idea in a secret notebook labeled "Transmutation of Species." This notebook contains Darwin's first "tree of life," in which he depicts life not as a hierarchical ranking of "higher" and "lower" forms (as Aristotle and other naturalists had claimed) but instead as a branching tree showing shared origins. The branches of the tree did not necessarily lead anywhere; they just spread. Instead of marching up a chain or ladder as Lamarck and other naturalists had suggested, Darwin's tree showed that species evolved; in some cases, one species could give rise to many species (as had occurred on the Galapagos Islands). Although Darwin would not publish his theory for 22 years, his "tree of life" would become a metaphor for his view of how species evolve.

While in the Galapagos, Darwin failed to record all the information he later needed to make full use of his data. For instance, Darwin understood—once his finches were properly identified after his return to England—that they were all closely related. However, while in the Galapagos, Darwin had paid them little mind, and because he had neglected to record the precise island from which each was taken, he could not reconstruct their probable relationships. Returning to the Pacific was out of the question, so he asked FitzRoy and other shipmates if he could borrow the Galapagos birds they had donated to the British Museum. Darwin received six sets of bird skins, and Gould's conclusion was strengthened: each island housed a different species of finch. Meanwhile, Thomas Bell, who had been identifying Darwin's reptiles, provided a parallel conclusion: each island of the Galapagos chain had produced its own distinct species of iguana.

Although Darwin's theory consumed much of his time, he began considering marriage. He listed the advantages of marrying ("constant companion and a friend in old age . . . better than a dog anyhow"), as well as the disadvantages ("less money for books," and "terrible loss of time"). On January 29, 1839—just five days after he was elected Fellow of the Royal Society—Darwin married his first cousin Emma Wedgwood at a ceremony officiated by Rev. John Allen Wedgwood (Emma's cousin) at the Church of St. Peter near the Wedgwood mansion. For the rest of his life, Charles referred to Emma as "my greatest blessing." The newlyweds moved into a house near Charles' brother Erasmus in London to start their family; they eventually had ten children, but only seven reached adulthood. Darwin was wealthy; after returning from his *Beagle* voyage, Darwin's father gave him stocks and a yearly allowance of £400, which was raised to £500 when Darwin married. Emma's dowry brought in another £5,000, and the Darwins inherited £45,000 when Charles' father died. Later in his life, Darwin—who invested primarily in railways and government bonds—had his son William Erasmus handle his finances. By 1881, Darwin's income had risen to £17,299, of which he spent £4,880 (he invested £10,218 and gave £3,000 to his children). Late that year, William informed his father that his estate was worth £282,000. The Darwins had no financial worries and never had to work.

While living in London, Darwin used a journal he originally wrote for his family as the basis for a book documenting his voyage aboard the *Beagle*. The book had a ponderous title: *Narrative of the Surveying Voyages of His Majesty's Ships* Adventure *and* Beagle, *between the Years 1826 and 1836 Describing their Examination of the Southern Shores of South America and The* Beagle's *Circumnavigation of the Globe in Three Volumes.* Darwin's book was Volume 3 of the set (another volume was written by FitzRoy). The volumes could be purchased separately, and Darwin's became a bestseller. When Henry Colburn, the publisher, reprinted Darwin's book, he gave it a grander title: *Journal of Researches into the Geology and Natural History of the Various Countries Visited by HMS* Beagle *Under the Command of Captain FitzRoy, R.N. from 1832 to 1836.* Darwin's book, retitled *The Voyage of the Beagle* at its third printing, was reprinted many times. This book, one of the world's great travel books, remains a steady seller.

Although Darwin, like Lyell, was praised as a scientist and writer, FitzRoy's volume—which included comments about geology and biblical history—was ridiculed, and Darwin's accomplishments soon relegated FitzRoy to a historical footnote. FitzRoy had taken Darwin around the world and made Darwin's discoveries possible, but he would regret his role in Darwin's achievements for the rest of his life.

Meanwhile, the Darwins soon tired of life in "dirty" London. In 1842, they paid the Rev. J. Drummond £2,020 for a large house near the village of Downe; Charles and Emma Darwin lived contentedly in Down House for the rest of their lives. Despite his subsequent fame, Darwin never again left England.

In 1842, Darwin developed the ideas in his "Transmutation" notebook into a thirty-five-page outline of "descent with modification" (as evolution was called in Darwin's day). Darwin discussed his idea with a few of his close friends, most notably Joseph Hooker, the founder of plant geography, and Charles Lyell, who along with James Hutton founded modern geology. After confiding to Hooker that he had discovered "the simple way which species become exquisitely adapted to various ends," Darwin likened his idea to "confessing a murder."

Darwin had witnessed the 1844 firestorm caused by Robert Chambers' *Vestiges of the Natural History of Creation*, and he was reluctant to announce his theory. Instead, he continued to do research that produced eight more books on topics that included insectivorous plants, earthworms, pigeons, barnacles, climbing plants, orchids, and plants' movements. Everywhere he looked, and regardless of what he studied, Darwin found evidence that supported his theory. In 1844, he expanded his 35-page outline into a 231-page (about 50,000 words) manuscript, which he stored in a hallway closet in Down House.

As an adult Darwin was chronically ill, prompting his friend Thomas Huxley to note that Darwin "might be anything if he had good health." Although Darwin continued to refuse to announce his idea, he set aside £400 for Emma to publish his expanded manuscript if he died unexpectedly.

On June 18, 1858, Darwin received a 28-page letter from British naturalist Alfred Russel Wallace, who was halfway through eight years of collecting specimens across the Malay Archipelago. (Wallace had written the letter in February, but it had taken four months to reach Darwin.) In that letter, Wallace described the same idea for evolution that Darwin had been secretly writing about. As Darwin noted after reading Wallace's letter, "I never saw a more striking coincidence. All my originality, whatever it may amount to, will be smashed."

After consulting with friends, Darwin outlined his ideas, and Darwin and Wallace's letters, along with part of Darwin's 1844 essay and an earlier letter from Darwin to Asa Gray, were read on July 1, 1858, at a meeting of the Linnean Society, a leading society of professional scientists in England. Darwin did not attend the meeting (he was mourning the death of his son Charles Waring Darwin, who had died two days earlier of scarlet fever), and Wallace did not know about the meeting. The presentation generated little interest among those who attended the meeting. The paper was then published under the impressive title *"On the Tendency of Species to Form Varieties; And on the Perpetuation of Varieties and Species by Natural Means of Selection* by Charles Darwin Esq., FRS, FLS, & FGS and Alfred Wallace Esq., communicated by Sir Charles Lyell, FRS, FLS, and J. D. Hooker Esq., MD, VPRS, FLS, &c." Darwin began writing a book describing his idea, and he finished the final chapter on March 19, 1859. Darwin understood the importance and potential impact of his idea—he noted, "It is no doubt the chief work of my life." Darwin's handwriting was poor, and publication of his book was delayed slightly because of questions raised by Ebenezer Norman, a schoolmaster at Downe who was the copyist of Darwin's manuscripts. On November 24, 1859, fully 22 years after Darwin had opened his secret "Transmutation of Species" notebook—John Murray Publishing of London (which had published all of Lyell's books) released Darwin's 502-page book *On the Origin of Species by Means of Natural Selection, Or The Preservation of Favored Races in the Struggle for Life.*

Murray printed 1,250 copies of *Origin of Species*, 139 of which were distributed as promotional copies. Publishers bought all of the remaining copies, each of which was sold for 15 shillings apiece. On the day of its publication, Murray asked Darwin if he wanted to make any changes before the second printing (Darwin changed about 7 percent of the text, which appeared as the second edition in 1860). Darwin's subsequent revisions were published in several languages that took Darwin's idea throughout the world. Darwin updated his theory and addressed critics' concerns by revising the book five times, and the final (i.e., sixth) edition was published on February 19, 1872. Darwin's idea became known as the "survival of the fittest," a phrase coined in 1863 by British philosopher and economist Herbert Spencer. Although neither this phrase nor the word "evolution" were in the first edition of *Origin of Species*, Darwin liked Spencer's phrase and believed that it was "more accurate" than his own explanation of natural selection. Darwin first used the phrase "survival of the fittest" in the title of Chapter 4 ("Natural Selection, or the Survival of the Fittest") of the sixth edition of *Origin*. During Darwin's lifetime, Murray sold approximately 25,000 copies of the English version of *Origin*. From 1859 to 1881, Darwin's books earned him an average of £465 per year; in 1871, these royalties constituted only about 6 percent of Darwin's total income.

Darwin's *Origin of Species* was not the first book about evolution, but it was—and still is—the most influential. Unlike Lamarck's *Philosophie Zoologique*, which was purely a theoretical book, Darwin's *Origin* was an overwhelming compendium of facts. Darwin stressed that his book was not a denial of God's existence, but it did challenge biblical literalism and remove humans from their pinnacle as the ultimate purpose of God's creation. Not surprisingly, *Origin* was condemned by many religious leaders, and William Whewell, Master of Trinity College at Cambridge, refused to allow it into the college library. However, many others praised Darwin's book. Darwin himself saw his idea as enlightening, as he noted in the book's famous final paragraph:

> *There is grandeur in this view of life, with its several powers, having been originally breathed into a few forms or into one; and that, whilst this planet has gone cycling on according to the fixed law of gravity, from so simple a beginning endless forms most beautiful and most wonderful have been, and are being, evolved.*

Unlike earlier explanations of evolution, Darwin's theory was supported by a huge amount of evidence and included a workable, coherent, and testable mechanism that did not require a deity, miracles, or arbitrary purpose. Just as Newton had done in *Principia*, Darwin included an enormous number of detailed observations to create "one long argument" for his theory.

The cornerstone of Darwin's theory is natural selection, the differential survival and reproduction of organisms. Natural selection produces adaptations, which are traits that enable organisms to survive the "struggle for existence." Darwin was convinced that just as domestic animals evolve through selective breeding (i.e., artificial selection), species in the wild evolve "by means of natural selection." Darwin explained natural selection in Chapter 4 of *Origin of Species* ("I have called this principle, by which each slight variation, if useful, is preserved, by the term Natural Selection"). For Darwin, natural selection was the force that constantly adjusts the traits of future generations by sorting hereditary variations. Darwin did not discuss the origin of life in *Origin of Species*, and referred to human evolution in one sentence that could be the understatement of the nineteenth century: "Light will be thrown on the origin of Man and his history." The implications of Darwin's theory were clear to contemporary readers:

> Darwin replaced the notion of a perfectly designed and benign world with one based on an unending, amoral struggle for existence.
>
> Darwin challenged prevailing Victorian ideas about progress and perfectibility with the notion that evolution causes change and adaptation but not necessarily progress and never perfection.
>
> Darwin's theory was theologically divisive, not because of what it implied about animal ancestry, but because it offered no purpose for humanity other than the production of fertile offspring.
>
> Darwin challenged the Providentially supervised creation of each species with the notion that all life—humans included—descended from a common ancestor. Humans are not special products of creation but products of evolution acting according to principles that act on other species.

Although Darwin knew that Wallace would have written an *Origin*-like book "if [he] had had my leisure," Wallace credited Darwin as the originator of the theory of evolution by natural selection. Darwin knew that his book would disturb many people; when he sent a copy of *Origin of Species* to Wallace late in 1859, he enclosed a note: "God knows what the public will think." However, Darwin had many defenders, most notably Harvard scientist (and evangelical Christian) Asa Gray in the United States, and Thomas Huxley in England. Throughout the uproar that followed the publication of his book, Darwin stayed at

Down House; he was interested in what was happening but stayed out of the fray.

After Darwin published *Origin*, many people continued to claim that humans are exempt from evolution. These claims inspired Darwin to write the two-volume work *The Descent of Man, and Selection in Relation to Sex*, which Murray published in 1870–1871 to address what Darwin called "the highest and most interesting problem for the naturalist." Part I of the book described evidence for human evolution (e.g., homologies and vestigial structures shared with apes). Part II described sexual selection, which Darwin used in Part III to explain human diversity and the origin of unique human traits. In this book, the first comprehensive theory of human evolution, Darwin emphasized that humans, like all other species, are subject to evolution by natural selection. Darwin made an accurate prediction about humans' origin:

> In each great region of the world the living mammals are closely related to the extinct species of the same region. It is, therefore, probable that Africa was formerly inhabited by extinct apes closely allied to the gorilla and chimpanzee; and as these two species are now man's nearest allies, it is somewhat more probable that our early progenitors lived on the African continent than elsewhere.

As Darwin noted in the book's last sentence, "Man with all his noble qualities . . . still bears in his bodily frame the indelible stamp of his lowly origin." Darwin knew that his conclusions would "be highly distasteful to many" but that "there can hardly be a doubt that we are descended from barbarians."

Although Darwin is best known for *Origin* and *The Descent of Man*, he also published several other important books, all of which supported his theory of evolution by natural selection. Darwin's two-volume *Variation of Plants and Animals Under Domestication* was meant to be the first part of Darwin's planned "big book" expanding on the "abstract" he had published as *Origin*. Despite its huge size, *Variation* sold well. Darwin's other books about orchids, insectivorous plants, climbing plants, and barnacles described numerous evolutionary adaptations and showed that slightly modified body parts could serve different functions in new environments (i.e., were adapted to "diversified places in the economy of nature"). Similarly, his *Expression of the Emotions in Man and Animals* discussed evolutionary aspects of emotions and behavior. In late 1881, Darwin's last major effort produced *Formation of Vegetable Mould, through the Action of*

Worms, with Observations on their Habits, a small, quirky book that discusses how small events over long periods of time can produce major results. While writing this book, Darwin visited Stonehenge to see how far worms might have buried the "Druidical stones." (Although Darwin also set up a "wormstone" in his backyard to study worms' movement of soil, the stone now at Down House was reconstructed by his son Horace's Cambridge Instrument Company in 1929.) In all, Darwin published 17 books in 21 volumes consisting of more than 9,000 pages of text and an additional 170 pages of preliminary matter.

Late in his life, Darwin noted, "I can indeed hardly see how anyone ought to wish Christianity to be true; for if so the plain language of the text seems to show that men who do not believe, and this would include my Father, Brother, and almost all my friends, will be everlastingly punished. And this is a damnable doctrine." Although Darwin argued against special creation and described himself as agnostic (a word coined by his friend Huxley), he did not publicly argue against religion; he denied that species have separate origins but did not deny the existence of God.

Late in 1877, in what was one of the proudest moments of his life, Darwin received an honorary doctorate from Cambridge University.

But his health continued to decline, and Darwin suspected that his death was imminent. After joining with Huxley to convince Queen Victoria to grant the financially strapped Wallace a lifelong government pension (£200 per year), Darwin made out a will in 1881 leaving £1,000 for his friends Hooker and Huxley "as a slight memorial of my lifelong affection and respect." In his seventies, Darwin wrote a short autobiography for his family. Darwin did not intend the autobiography for publication, but a censored version of the manuscript was published in 1887 by Francis Darwin as *The Autobiography of Charles Darwin*. Years later, Darwin's granddaughter Nora published the entire original manuscript.

By early March 1882, Darwin's health began to fail in earnest. He took his last stroll along his beloved Sandwalk on March 7, after which he became increasingly ill. On April 19, 1882, Darwin told Emma to "remember what a good wife you have been" and that he "was not the least afraid to die." Later that afternoon, at age 73, Charles Darwin died in his upstairs bedroom at Down House.

The world noted the passage of Darwin and his towering intellect. A newspaper in Vienna noted, "Humanity has suffered a great loss.... [O]ur century is Darwin's century," and in Paris, the editors of *France* proclaimed that Darwin's work was "an epic—the great poem of the genesis of the universe, one of the grandest that ever proceeded from a human brain." In London, the editors of the *Times* wrote, "One must seek back to Newton, or even Copernicus, to find a man whose influence on human thought . . . has been as radical as that of the naturalist who has just died. . . . Mr. Darwin will in all the future stand out as one of the giants in scientific thought and scientific investigation."

On the afternoon of April 25, Darwin's body was carried from Down House in a horse-drawn hearse. At noon the following day at a packed funeral service attended by Britain's leading politicians, clergy, and scientists (Hooker, Wallace, Huxley, and Darwin's neighbor John Lubbock were among the pallbearers), the choir sang an anthem specially composed by the abbey's organist Sir John Bridge (from Proverbs 3:13–17) to exalt Darwin's life of thought: "Happy is the man that findeth wisdom and getteth understanding. . . ." Darwin was buried in a white-oak coffin (made by Downe undertaker and builder John Lewis) in the northeast corner of the nave of London's Westminster Abbey beside astronomer John Herschel and near his friend Sir Charles Lyell. As Darwin's body was lowered into the abbey's floor, the choir sang, *His Body Is Buried in Peace, But His Name Liveth Evermore*. Darwin was the first and only naturalist to be buried in Westminster Abbey.

Darwin has been memorialized in many ways. More than 200 places, plants, animals, and awards are named for him. A *darwin* is a SI unit of evolutionary change (proposed by J. B. S. Haldane) equal to a rate of change of 0.1 percent per thousand years. (This unit is seldom used because there is little agreement about what should be measured to compute the rate of change.) Darwin's time in Edinburgh is commemorated by a plaque at the Royal Museum of Scotland. Darwin College of Cambridge (est. 1964) honors Darwin. In 2000, Darwin's image replaced fellow Victorian Charles Dickens on the British £10 note.

ERASMUS DARWIN (1731–1802)

Would it be too bold to imagine, that all warm-blooded animals have arisen from one-living filament?

Charles Darwin's eclectic grandfather, Erasmus Darwin, was born on December 12, 1731, in Elston, England. He was the seventh child of lawyer Robert Darwin (1682–1754) and Elizabeth Hill Darwin (1702–1797). Darwin's baptism was a feast that included a special beer bottled in his honor. Two unopened bottles, and the feast's menu, survive.

Darwin, aided by a scholarship of £16 per year, attended Cambridge from 1750–1754 and obtained a medical degree from Edinburgh Medical School in 1756. He began practicing medicine at Lichfield,

Staffordshire, and became rich by attending to wealthy clients. Darwin promoted education for women and despised slavery. An avid fossil collector who admired geologist James Hutton, he declined King George III's invitation to be Royal Physician.

In the mid-1860s, Erasmus helped found the Lunar Society of Birmingham, an informal group of industrialists and philosophers who "were united by a common love of science, which we thought sufficient to bring together persons of all distinctions [including] Christians . . . and Heathens." The Lunar Society, which met at each full moon, was an important organization for England's intellectuals in the second half of the eighteenth century. Darwin's friends and associates in the Lunar Society included Joseph Priestley (1733–1804; the preacher who discovered oxygen), flamboyant industrialist Matthew Boulton (1728–1809), Boulton's partner James Watt (1736–1819; inventor of the steam engine), and ambitious potter Josiah Wedgwood (1730–1795; Charles Darwin's other grandfather whose company—Josiah Wedgwood & Sons, Limited—remains a thriving company today). The Lunar Men blended science, commerce, and art, and in the process, helped start the Industrial Revolution. In 1783, Erasmus also founded the influential Derby Philosophical Society. Some of Erasmus' work inspired Mary Shelley to write *Frankenstein* (Erasmus is mentioned in the preface to the 1818 edition and in the introduction to the 1831 edition of the novel).

Darwin, who translated parts of Linnaeus' books, invented, or helped invent, a horizontal windmill, a canal lift for barges, a copying machine, and a variety of weather-related gadgets. Erasmus never patented any of his inventions (he believed that doing so would damage his reputation as a physician), and he encouraged others to modify and use his inventions.

Darwin was also an accomplished poet, and his poems were admired by Shelley, Coleridge, and Wordsworth. For example, Coleridge—who coined the term *darwinizing* to refer to Erasmus' speculations—suggested that Erasmus possessed "a greater range of knowledge than any other man in Europe, and is the most inventive of philosophical men." One of Darwin's most famous books of poetry, *The Botanic Garden* (1789), speculated about cosmological theories and noted that "the ingenious theory of Dr. Hutton" implied an eternal nature of Earth. Darwin's practical *Phytologia, or the Philosophy of Agriculture and Gardening* (1800) claimed that plants have senses and volition. Darwin speculated that electricity is the basis for nerve impulses, and that there is a "resemblance between the action of the human soul and that of electricity."

Erasmus Darwin married twice and had 14 children. His marriage in 1757 to Mary "Polly" Howard (1740–1770) produced a daughter and four sons, the third of which—Robert Waring Darwin (1766–1848)—was Charles Darwin's father. When Mary died in 1770, Erasmus fathered two illegitimate daughters with his 17-year-old mistress Mary Parker, after which he married widow Elizabeth Pole in 1781 and moved to her home near Derby. Erasmus and Elizabeth had four sons and three daughters. Violetta, the eldest of these daughters, was the mother of eugenics crusader Francis Galton.

In 1794, Erasmus published *Zoonomia, Or, the Laws of Organic Life*, his most important book. *Zoonomia* includes a system of pathology and a section that anticipated the views of Jean-Baptiste Lamarck. In *The Temple of Nature* (published posthumously in 1803), Erasmus used rhymed couplets to describe a gradual progress of life toward higher levels of complexity and greater mental powers. Erasmus never thought of natural selection, but he did suggest that all species have a common ancestor. Darwin also suggested that species' survival was governed by "laws of nature" rather than divine authority and that new species arise because of competition and sexual selection. However, like others before him, Darwin had little evidence to support his claims; his arguments were not convincing, and he could not explain adaptations. Nevertheless, many people recognized his contributions. For example, the preface to George Bernard Shaw's *Back to Methuselah* describes Erasmus as one of the originators of evolutionary theory.

Charles Darwin never met his eccentric grandfather (Erasmus died seven years before Charles was born) and was not overly impressed by his scientific claims. Nevertheless, Charles admired Erasmus, and when Charles was 70 years old he wrote Erasmus' biography, *The Life of Erasmus Darwin*. Charles' daughter Henrietta edited the book and removed parts of the text (about 16% of the total) that she considered too salacious for the Victorian audience. In his later years, Erasmus became a hedonist and was enormously fat; he had to carve a semicircle in his dining-room table to accommodate his expanding girth.

On March 25, 1802, Darwin moved to Breadsall Priory, just five miles north of Derby. He had inherited a house there from his son Erasmus, Jr. (1759–1799), who at age 40 had drowned—probably a suicide—in the Derwent River on December 29, 1799. Two weeks after moving to Breadsall Priory, Erasmus became ill, and a week later—on April 18, 1802—he died suddenly and painlessly. He was buried beside his son, Erasmus, Jr., on April 24 at All Saints Church in Breadsall; his wife was also buried there in February 1832. When the church was renovated in 1877, Darwin's coffin was opened, and his granddaughter Elizabeth Whelen described his remains as being "in perfect preservation. He was dressed in a purple velvet dressing-gown and his features unchanged." Today, Erasmus rests in the center of the nave of All Saints Church in Breadsall, and a monument erected by his widow honoring Erasmus' "zealous benevolence" adorns the church's south wall. Darwin is also commemorated with a medallion on the Exeter Bridge.

You can visit the Erasmus Darwin House on Beacon Street in Lichfield. The impressive house, which was opened to the public in 1999, is a tourist attraction and research center and includes an herb garden and a reconstruction of Darwin's medical office.

WILLIAM DEMBSKI (b. 1960)

Convinced Darwinists ... need to block the design inference whenever it threatens to implicate God. Once this line of defense is breached, Darwinism is dead.

William Albert Dembski was born July 18, 1960, in Chicago, Illinois. Dembski earned a B.A. in psychology in 1981 and an M.S. in statistics in 1983 from the University of Illinois at Chicago (UIC), a Ph.D. in mathematics from the University of Chicago in 1988, M.A. and Ph.D. degrees in philosophy from UIC in 1993 and 1996, respectively, and a Masters of Divinity from Princeton Theological Seminary in 1996. He also completed postdoctoral work at MIT (mathematics), the University of Chicago (physics) and Princeton University (computer science). He was awarded the Templeton Foundation's Book Prize ($100,000) for 2000–2001 for his writings on information theory.

Dembski did not question evolution as he was growing up but became increasingly skeptical of the power of natural selection by the time he began college. In 1988, he attended a conference about randomness that noted that patterns are inevitably discovered within what initially appears random. Dembski concluded that randomness was "always a provisional designation until we found the pattern or design in it," a conclusion that prompted him to begin studying design in nature.

In 1992, Dembski met philosopher Stephen Meyer. Dembski, then in graduate school at the University of Chicago, was "doing the design work on the side," and had attracted Meyer's attention. Meyer and Dembski joined with other intelligent design (ID) proponents, including Phillip Johnson, Michael Behe, and Jonathan Wells, all of whom would soon become fellows (Meyer as director) of the Center for the

Renewal of Science and Culture (later renamed the Center for Science and Culture) at the Discovery Institute. The duo of Behe and Dembski provided ID proponents with what they considered the necessary one-two punch: Behe's *Darwin's Black Box* (1996) purported to demonstrate the "irreducible complexity" of the basic unit of all life, the cell, while Dembski's work claimed evidence for the impossibility of life arising without the aid of a designer.

In 1998, Dembski's *The Design Inference: Eliminating Chance through Small Probabilities* introduced the use of an "explanatory filter" to identify design in nature. Dembski proposed that all patterns in nature have only three possible causes: regularity (the action of natural laws, e.g., gravity), chance, and design. The filter works as a simple flowchart: if the pattern can be explained by natural laws, the pattern is explained, otherwise move onto chance as an explanation, and then, finally, design.

Distinguishing patterns that are simply unlikely from those possible only through design—the heart of the matter for Dembski—requires identification of patterns that are not only unlikely (i.e., complex) but also demonstrate "specified complexity." For example, any particular sequence of heads and tails when flipping a coin 100 times is equally unlikely, and, therefore, any sequence so obtained would not require an explanation invoking anything beyond chance. Improbability alone, therefore, is not sufficient for identifying design. However, a specific sequence—say 100 consecutive heads—is extraordinarily unlikely, and, if encountered, would require an explanation that involves more than just chance. The *irreducible* complexity Behe claimed to have demonstrated was exactly the *specified* complexity Dembski required as evidence for design. Using a complicated formula that includes the number of particles in the universe and a physical constant known as Planck time, Dembski proposed that patterns less likely than one chance in 0.5×10^{150} are evidence for design.

Dembski's *Intelligent Design: The Bridge Between Science and Theology* (1999) continued his exploration of specified complexity, casting the search for design within the framework of information theory. In 2002, Dembski's *No Free Lunch: Why Specified Complexity Cannot Be Purchased Without Intelligence* extended his search for design through application of "No Free Lunch" (NFL) theorems. NFL theorems, in general, compare the overall performance of alternative "search algorithms" for accomplishing particular outcomes (e.g., the best route to take through a landscape to reach a desired objective). Dembski concluded that natural selection cannot perform better than random chance in generating specified complexity. Dembski's efforts, consequently, moved quickly from merely suggesting how design can be identified to a refutation of evolution.

Critics of Dembski's work noted that even though *Design Inference* passed inspection by the Cambridge University Press editorial board, none of Dembski's work had received standard peer evaluation. Dembski has claimed that he prefers to disseminate his work in book form because it is a faster process (Dembski has authored or coauthored an average of one book per year). Despite his training in mathematics, Dembski has been criticized harshly for his "mathematism," which relies on "pseudo-mathematical jargon" to cloak nonscientific ideas in scientific wrappings. Accusations of the misuse of probability, and, most recently, misapplication of the NFL theorems, have generated negative reviews (e.g., David Wolpert, who in part developed the NFL theorems, has described Dembski's work as "written in jello"). Dembski's proposed "filter" for identifying design has also been criticized as being inappropriate for ruling out the action of adaptive evolution (i.e., natural selection is not goal-directed) and for equating the action of natural selection with chance. ID proponents have, however, portrayed Dembski's work as a rigorous refutation of evolution (e.g., University of Texas philosopher and Discovery Institute fellow Robert Koons has called Dembski "the Isaac Newton of information theory").

Dembski's professional career started as a fellow with the Discovery Institute in 1996, where he is now a senior fellow. In 1999, Dembski joined Baylor University as Associate Research Professor and soon thereafter was appointed by Baylor's president Robert Sloan to head the newly created Michael Polanyi Center (MPC) for Complexity, Information, and Design. The secretive nature of the creation of the MPC caused dissent among Baylor's faculty, especially when Dembski's agenda for the study of ID became apparent. In response, Sloan convened a committee of outside experts that ultimately recommended folding the MPC into an existing university entity. Dembski publicly declared the committee's findings a victory for ID over its opponents ("[they] have met their Waterloo"). When Sloan asked Dembski to retract the comments, Dembski refused, and Sloan removed Dembski as director of the MPC. Dembski, who had a multiyear contract with Baylor, remained at the school for five more years, during which time he worked off-campus on various books. Dembski later remarked, "In a sense, Baylor did me a favor. I had a five-year sabbatical."

In 2005, Dembski moved briefly to the Southern Baptist Theological Seminary in Louisville, Kentucky, where he established the seminary's Center for Science and Theology. Less than a year later, he moved to the

Southwestern Baptist Theological Seminary in Fort Worth, Texas, accepting the position of Research Professor in Philosophy. (Kurt Wise, formerly of Bryan College, replaced Dembski in Louisville.) In 2001, Dembski established the International Society for Complexity, Information, and Design, "a cross-disciplinary professional society that investigates complex systems apart from external programmatic constraints like materialism, naturalism, or reductionism."

Dembski also advocated the "wedge" strategy developed in the 1990s at the Discovery Institute. He later proposed that the strategy be morphed into a "vise" strategy (subtitled "Squeezing the Truth out of Darwinists") that would "make clear to those reading or listening to the Darwinists' testimonies that their defense of evolution and opposition to ID are prejudicial, self-contradictory, ideologically driven, and above all insupportable on the basis of the underlying science." Dembski is remarkably antiscience, claiming that "the scientific picture of the world championed since the Enlightenment is not just wrong, but massively wrong."

In the wake of the anti-ID decision in the *Kitzmiller v. Dover Area School District* case, Dembski posted on his blog an animation (that he helped create) showing Dover presiding judge John Jones as a flatulent puppet of the ACLU and evolutionary biologists. (Dembski originally agreed to testify in the Dover trial but withdrew before the trial started.) Dembski was also in "constant correspondence" with conservative commentator Ann Coulter during development of her confrontational book, *Godless: The Church of Liberalism* (2006), which claims that evolution is not science (e.g., it is no better supported than Coulter's mocking "Flatulent Raccoon Theory" for the origin of life) and represents part of the "religion" of liberalism.

DISCOVERY INSTITUTE (Est. 1990)

The Discovery Institute was founded as a Seattle branch of the Hudson Institute, a conservative think tank based in Indianapolis, Indiana. Bruce Chapman (b. 1940), a former Secretary of State for the State of Washington (who also served as Director of the U.S. Census Bureau in the early 1980s and as Deputy Assistant to President Ronald Reagan from 1983 to 1985), joined the Hudson Institute in the late 1980s and agreed to return to the state of Washington to open the Seattle office.

The Discovery Institute originally focused on regional economic, communications, and transportation policy. However, intelligent design (ID) became a major emphasis of the Institute after Chapman met Stephen Meyer in 1994. At that time, Meyer began working with Phillip Johnson, William Dembski, and Michael Behe to coalesce the ID movement. Chapman likely realized that the perceived conflict between science and religion, in the guise of ID, could attract wealthy donors to his financially struggling organization that was now independent of the Hudson Institute.

Meyer had tutored the children of Howard Ahmanson, a wealthy conservative who had inherited a fortune from his banker father, and he set up a meeting among Ahmanson, Chapman, and himself. Ahmanson, who supported "the total integration of biblical law into our lives," provided $750,000 over three years to found the Center for the Renewal of Science and Culture (CRSC) within the Discovery Institute. The CSRC became the research center for ID, and Meyer, Behe, Dembski, and Jonathan Wells were named fellows of the Center. Meyer became director of the CRSC, and Johnson served as advisor.

Johnson, a Berkeley law professor who concluded in *Darwin on Trial* (1991) that evolution was an unsupported theory, wanted to undermine what he saw as the dominant and pernicious materialism that excluded God as an explanation for nature. Johnson and others developed a multiyear plan—the "wedge strategy"—that used "atheistic evolution" to "replace materialistic explanations with the theistic understanding that nature and human beings are created by God." The strategy was to be kept secret, but the document detailing the plan was discovered when sent out for photocopying and was anonymously posted on the Internet. The CRSC (renamed the Center for Science and Culture to reduce the aura of a religion-based agenda) initially denied authorship of the "wedge document," although it eventually acknowledged that it had originated from the Discovery Institute as an "early fundraising proposal."

Out of the controversy surrounding public scrutiny of the wedge document, and coincident with the Kansas Board of Education's decision to stop requiring the teaching of evolution in public schools, the CSC morphed its wedge strategy into "teach the controversy." This approach does not exclude evolution from science curricula but does portray it as a "theory in crisis." For the teaching of evolution and questions of origins to be impartial, therefore, adequate coverage of the problems and limitations of the theory are required, which in turn allows inclusion of alternative explanations such as intelligent design. The CSC claims that these efforts promote "academic freedom" so teachers can overcome the "Darwinian fundamentalists" who are "waging a malicious campaign to demonize and blacklist anyone who disagrees with them." Meyer also claims that the "teach the contro-versy" proposal is federally mandated, appealing to *Edwards v. Aguillard* (1987), which he interprets as requiring an equal hearing of all alternative scientific theories. (This strategy actually originated with Wendall Bird of the Institute for Creation Research in the 1980s.) CSC claims that ID is neutral as to who or what the designer is, and therefore ID is a scientific explanation that should be heard. The Discovery Institute claims that it rejects efforts to prevent the teaching of evolu-tion, or to require the teaching of any particular theory, including ID. For example, the Discovery Institute publicly disagreed with the "mis-guided policies" of the Dover, Pennsylvania, School Board to include ID in ninth-grade biology. However, CSC fellows testified on behalf of the defendants, and the Discovery Institute condemned the conclu-sions of the presiding judge, John Jones.

CSC helps sympathetic state legislators craft legislation advocating the "teach the controversy" approach. These efforts reached the federal level in 2001 when Johnson wrote what become known as the "Santorum Amendment." This proposed amendment to the 2001 Elementary and Secondary Education Act Authorization Bill (later known as the No Child Left Behind Act) was introduced by Rick Santorum, then a U.S. Senator from Pennsylvania. The amendment proposed that biology instruction should discuss "why [evolution] gen-erates so much continuing controversy," a clear reflection of the CSC's main strategy. The amendment was deleted from the final bill, although the Discovery Institute claims that "teaching the controversy" became federal law.

Although the Discovery Institute promotes itself as the research home for ID (e.g., Chapman announced in 2006 that the Institute had "put over $4 million toward scientific and academic research into evo-lution and intelligent design" since 1996), critics have noted that the Institute has not produced peer-reviewed publications. In 2001, the Discovery Institute launched "A Scientific Dissent from Darwinism," which asked those holding a Ph.D. in science, engineering, mathemat-ics, or computer science to endorse the statement, "We are skeptical of claims for the ability of random mutation and natural selection to account for the complexity of life." As of 2007, more than 600 signatures had been collected. However, *The New York Times* and other organiza-tions reported that most individuals who endorsed this statement are neither biologists nor experts in evolutionary biology. (The National Center for Science Education started the "Project Steve" parody, which asks Ph.D.-level scientists named "Steve"—which should be about 1 percent of all scientists and was chosen in honor of the late Stephen

Gould—to sign a letter in support of evolution. As of 2007, more than 800 names had been collected supporting "Project Steve." Also, at the time of the Dover trial in 2005, the "A Scientific Support for Darwinism" project collected 7,733 signatures from scientists worldwide in four days.) The Discovery Institute also provided a viewer's guide titled *Getting the Facts Straight* to accompany PBS's 2002 series on evolution. The guide condemns the documentary as a "misuse of taxpayer money to organize and promote a controversial political action agenda."

The Discovery Institute also studies transportation issues in the Pacific Northwest (the Cascadia project), the role of technology in improving the quality of life (the Technology and Democracy Project), and national fiscal and monetary policy. The Institute is well funded, with an annual budget exceeding $4 million (including $1 million a year for ten years from the Bill and Melinda Gates Foundation to support the Cascadia project). ID, however, is viewed by Chapman as his organization's "number one project." The Discovery Institute plans to release a textbook titled *Explore Evolution* that purports to discuss alleged "scientific controversies" associated with evolution.

JAMES WATSON (b. 1928)

The book of the DNA sequence would in time be regarded as more relevant to human life than the Bible. It tells us who we are. I've never read the Bible, so I'm not sure I've missed much.

James Dewey Watson was born on April 6, 1928, in Chicago, Illinois. Watson attributed his self-confidence, libertarian spirit, and trademark frankness—what some would call impertinence—to his parents' supportive influence. Above all, inquiry and knowledge were valued in his home; as Watson recalled, "I was sort of trained to get pleasure from understanding the world around me, not from material things." Watson's father was an avid birdwatcher, and father and son took regular walks to look for birds. At 15 years old, Watson entered the University of Chicago, intent on a career in ornithology. However, he became interested in genetics, and after graduating in 1947 with a B.S. in zoology, decided to study genetics in graduate school.

In 1947, Watson earned a fellowship to study at Indiana University and started working in microbiologist Salvador Luria's lab. Luria was part of the "Phage Group," which included Max Delbrück and Alfred Hershey (both of whom, along with Luria, would share a Nobel Prize in 1969). These biologists used bacteriophage (viruses that infect bacteria) to study inheritance. The group met regularly at Cold Spring

Harbor Laboratory, and Luria included Watson in these meetings. The Phage Group was convinced that DNA was the molecule of heredity (rather than proteins), and Watson later credited the early opportunity to work closely with such pioneers as crucial to his ultimate success. Watson finished his Ph.D. in 1950, keen on studying genetics. This meant analyzing the structure of DNA, and he accepted a postdoctoral position in Copenhagen.

Although his postdoc was a disappointment (his new advisor was not studying DNA's structure), Watson heard a presentation by Maurice Wilkins of King's College, London, which described how x-ray diffraction could be used to study the structure of DNA. This was exactly the research Watson wanted to do, and he transferred to the Cavendish Laboratory at Cambridge, the leading center for x-ray diffraction studies. Soon after arriving in Cambridge in 1951, Watson befriended graduate student Francis Crick, who was interested in the same questions.

Watson and Crick were assigned a project on the structure of proteins but they remained obsessed with DNA. Watson in particular was driven to solve the mystery of DNA as quickly as possible because he knew that Linus Pauling in the United States was working on the same question. Following Pauling's lead, Watson and Crick not only approached the task through theory and experiment, they also built elaborate models out of wire, metal, and cardboard. But it was not until Wilkins let the two glimpse results produced by Wilkins' estranged colleague Rosalind Franklin that they identified the simple yet elegant structure of the DNA double helix. Watson and Crick published their findings in a short paper in *Nature* in 1953, followed by three others over the next year that detailed their work. This research earned Watson, Crick, and Wilkins the Nobel Prize for Physiology or Medicine in 1962.

In 1953, Watson moved to the California Institute of Technology to perform similar studies of RNA. He returned to the Cavendish Laboratory briefly in 1955 before settling in at Harvard in 1956, attaining the rank of professor five years later. Watson continued to study RNA and ribosomes, forming a long-term collaboration with Walter Gilbert (winner of the Nobel Prize in 1980). Watson's bluntness also earned him a dubious reputation among his colleagues, especially from organismal biologists, whom Watson labeled "stamp collectors" left behind by the new era of molecular biology. Noted Harvard ecologist Edward O. Wilson in turn later declared Watson "the most unpleasant human being I had ever met."

While at Harvard, Watson published *The Double Helix: A Personal Account of the Discovery of the Structure of DNA* (1968), a frank account of the race to determine the structure of DNA. The book was considered so inflammatory (including by Crick) that Harvard University Press rescinded its agreement to publish it. (The book was eventually released by Atheneum.) Watson was also criticized for his harsh treatment of Rosalind Franklin, who had done the x-ray diffraction work critical to Watson's and Crick's conclusions. (Franklin had died of ovarian cancer in 1958.) Regardless, the book, which became a bestseller, is included in the Modern Library's list of 100 best nonfiction books of the twentieth century, and it was made into a 1987 movie starring Jeff Goldblum as Watson. Watson also helped write the textbook *The Molecular Biology of the Gene* (1965), which provided a contemporary discussion of molecular biology and set a new standard for textbook design.

In 1968, Watson became director of the Cold Spring Harbor Laboratory (CSHL) and also married his wife, Elizabeth, with whom he eventually had two sons. Watson left Harvard in 1976 for permanent residence at CSHL, became president of CSHL in 1994, and then chancellor in 2004 when CSHL became a graduate-degree-granting institution with the opening of the Watson School of Biological Sciences. Watson is credited with rescuing CSHL, primarily by emphasizing cancer research that attracted large grants, when the Carnegie Institution removed its financial support in the 1960s. From 1988 to 1992, Watson also served as the first director of the Human Genome Project (HGP). At a news conference early in his tenure with the HGP, he made an on-the-fly decision to devote 3 percent (later increased to 5 percent) of the program's budget to study ethical and social implications of the project. This was hailed as a major endorsement of the need for bioethical analysis in science, especially genetics. Watson stepped down (he says he was "fired") as director of the HGP over disagreements with the director of the National Institutes of Health about granting patents for DNA sequences of unknown function.

In *DNA: The Secret of Life* (2003), Watson supported the development of genetically modified foods, discussed the need for confidentiality of personal genetic information and rejected the current system of patenting genetic information. Watson also endorsed "human germline engineering," noting that "we have really got to worry that a genetic underclass exists." Later, he became even more brash, commenting, "If you are really stupid, I would call that a disease. . . . So I'd like to get rid of that, to help the lower 10 percent." Watson also hopes that genetic engineering will make people more compassionate: "We'll understand why people can't do certain things. Instead of asking a child to shape up, we'll stop having unrealistic expectations."

Scientists, especially cognitive scientists, have objected to Watson's implied genetic determinism of a complex trait like intelligence, and religious leaders have warned of the dangers of "playing God." Watson's militant atheism—"The biggest advantage to believing in God is you don't have to understand anything"—brought condemnations by believers. Bioethicists highlight both the ethical problems of designing children and the sinister nature of past eugenic movements. Undaunted, Watson urged a group of 1,000 German scientists to "put Hitler behind us" and use genetics to improve humans. Ironically, one of Watson's sons suffers from a serious cognitive disorder, and Watson was chancellor of an institution—Cold Spring Harbor Laboratory (CSHL)—that, through the Eugenics Record Office, led twentieth-century eugenics efforts in the United States. In 2005, Watson agreed to have his genome sequenced and made public; Watson's sequence is now accessible through a National Institutes of Health database. Watson resigned his position at CSHL in late October 2007 after he provoked widespread outrage by claiming that Africans are intellectually inferior to Westerners.

FRANCIS CRICK (1916–2004)

The age of the Earth is now established beyond any reasonable doubt as very great, yet in the United States millions of Fundamentalists still stoutly defend the naive view that it is relatively short, an opinion deduced from reading the Christian Bible too literally. They also usually deny that animals and plants have evolved and changed radically over such long periods, although this is equally well established. This gives one little confidence that what they have to say about the process of natural selection is likely to be unbiased, since their views are predetermined by a slavish adherence to religious dogmas.

Francis Harry Compton Crick was born on June 8, 1916, to Harry Compton Crick and Annie Elizabeth Crick. Francis's grandfather was an avid amateur naturalist and in early 1882 discovered a water beetle with a small mollusk attached to its leg. He was so enthralled by this discovery that he wrote to the greatest naturalist of the day, Charles Darwin. Darwin was also impressed, having earlier proposed that mollusks may be dispersed in this way, and had the specimen sent to him. Darwin published the findings in *Nature* in April of 1882, only 13 days before he died.

By the time Francis was born, World War I had pushed England into an economic slump that caused the family business to fail, and the Cricks were forced to move to London. In 1930, Francis won a scholarship to the Mill Hill boarding school in London, where he exhibited an affinity for science. Even so, he was denied entry to Cambridge and Oxford and instead entered University College, London, in 1934, graduating three years later with a B.S. in physics. He immediately started working toward a Ph.D. in physics at University College, but the outbreak of World War II forced him to suspend his graduate education. Crick joined the British Admiralty Research Laboratory and remained there until 1947, developing mines for the military. During this period, he married his first wife, Ruth, with whom he had a son, Michael. The Cricks divorced in 1947.

A reading of physicist Edwin Schrödinger's book *What Is Life? The Physical Aspects of the Living Cell* (1944) led Crick away from particle physics and into biology. He transferred to the Strangeways Research Laboratory at Cambridge, where he learned x-ray diffraction, a tool that would be crucial for identifying the structure of DNA. In 1949, Crick moved again, this time to the Medical Research Council Unit for Molecular Biology at the Cavendish Laboratory at Cambridge University. The Cavendish Laboratory was famous for research in physics (James Clerk Maxwell was its first director), and in the late 1930s had added molecular biology as an area of study. It was here that Crick's interest in genetics began. Crick married his second wife, Odile, in 1949 and had two daughters with her.

American biologist James Watson joined the Cavendish Laboratory in 1951. Watson was keen on using x-ray crystallography to investigate the structure of DNA, and Crick and Watson immediately started working together. Combining recent research by others (including data that may have been improperly obtained from ex-colleague Rosalind Franklin) with their own work, Watson and Crick built models (using both mathematics and cardboard) showing the possible configuration of the DNA molecule. Ultimately, they determined the structure of the now well-known double helix and published their results in *Nature* in 1953. The following year, Crick earned his Ph.D. for research on helical protein structures. Crick, Watson, and colleague Maurice Wilkins (a physicist who had worked with Rosalind Franklin on x-ray diffraction studies of DNA) shared the Nobel Prize for Physiology or Medicine in 1962 for their work on DNA.

In 1957, Crick began studying mutations in viruses with biologist Sydney Brenner (who would win a Nobel Prize in 2002). These experiments eventually linked nucleotide sequences in DNA with amino acid sequences in proteins. Work by others showing the involvement of RNA in protein synthesis allowed Crick to elucidate the triplet nature of the genetic code and to propose the "central dogma" of molecular biology: DNA → RNA → protein. In 1962, Crick became codirector of Cambridge University's Molecular Biology Laboratory with Brenner.

Crick's understanding of molecular biology—what he originally referred to as "the borderline between living and the nonliving"— encouraged him to consider how life may have arisen. Crick concluded that the production of life from molecules that were not living was an extremely rare event and unlikely to have happened on Earth. This idea was expanded in Crick's 1981 book *Life Itself: Its Origin and Nature*, which invoked the concept of "directed panspermia"—the idea that life had arisen elsewhere in the universe and had been dispersed to Earth by intelligent alien life forms. By the 1990s, however, discovery of enzymatic RNA had fostered development of the "RNA world" hypothesis for the origin of life on Earth, and Crick acknowledged that he had underestimated the probability of life arising on Earth without outside influence. Regardless, Crick's statements (e.g., "The probability of life originating at random is so utterly minuscule as to make it absurd") have been cited repeatedly by antievolutionists as evidence that science has proven the need for a supernatural explanation for life. The rise of intelligent design (ID) provided yet another opportunity for Crick's ideas, particularly panspermia, to be misused. For example, during the *Kitzmiller et al. v. Dover Area School District* trial, ID advocate Michael Behe used Crick's writings to suggest that scientists naturally believe the universe is designed.

When he was 12 years old, Crick stopped attending church, and he remained an atheist throughout his life. In his 1967 book *Of Molecules and Men*, Crick confronted the concept of "vitalism"—that is, the idea that life cannot be explained entirely by natural phenomena open to scientific inquiry. He opposed mixing religion and science and resigned as fellow of Churchill College, Cambridge, when the college decided to build a chapel on campus. His refusal to allow religion to dictate how science is taught also spurred him to join over 70 other Nobel laureates in signing an amicus brief stating " 'creation-science' simply has no place in the public-school science classroom" for the *Edwards v. Aguillard* case in 1987.

Crick relied on what he called the "gossip test"—the notion that what you talk about in everyday conversation likely reflects your primary interests—to determine his main scientific interests: molecular biology and the workings of the brain. Having successfully studied the first area for several decades, he decided in 1977 to investigate the second by accepting the position of Kieckhefer Distinguished Research Professor at the Salk Institute of Biology in La Jolla, California, where he would remain for the remainder of his career. In 1994, he published *The Astonishing Hypothesis: The Scientific Search for the Soul*, which concluded that self-awareness results from interactions of the multitude of neurons in the brain, and that the human "soul" is not an entity separate from the understandable workings of the brain.

Francis Crick died on July 28, 2004, of colon cancer.

WILLIAM PALEY (1743–1805)

There cannot be design without a designer.

William Paley was born in Peterborough, England, in July 1743. He graduated from Cambridge University first in his class in 1763. Paley became a deacon in 1765, was appointed assistant curate in Greenwich and taught at Cambridge for 10 years. He was ordained in 1767 (after earlier earning an M.A.), and the remainder of his clerical career included successively more important positions within the Anglican Church. Paley opposed slavery and advocated prison reform, and as a philosopher he was a utilitarian, believing that humans act morally to increase their overall level of happiness. In 1776, Paley married Jane Hewitt, with whom he had eight children.

Paley was a popular preacher and one of England's most important theologians of his generation. He published his Cambridge lectures in *The Principles of Moral and Political Philosophy* (1785), which outlined his utilitarianism and was used as a textbook at Cambridge for many years. This book was followed by *A View of the Evidences of Christianity* (1794), a response to David Hume's skepticism of religion and, in particular, Hume's dismissal of miracles. But Paley's best-known book, and his last, was *Natural Theology; or, Evidences of the Existence and Attributes of the Deity, Collected from the Appearances of Nature* (1802).

In *Natural Theology*, Paley—one of the most admired clerics in the English-speaking world—argued that God could be understood by studying the natural world. *Natural Theology* begins with the famous metaphor of "God as watchmaker":

> In crossing a heath, suppose I pitched my foot against a stone, and were asked how the stone came to be there; I might possibly answer, that, for any thing I knew to the contrary, it had lain there for ever: nor would it perhaps be very easy to show the absurdity of this answer. But suppose I had found a watch upon the ground, and it should be inquired how

the watch happened to be in that place; I should hardly think of the answer which I had before given, that, for any thing I knew, the watch might have always been there.

Paley argued that the only rational conclusion was that the watch had a designer. Much of *Natural Theology* discusses examples of purported design in "a happy world" that teems "with delighted existence," with many drawn from Paley's own observations and likely to be familiar—and therefore persuasive—to readers. Paley's designer was his watchmaking God.

Charles Darwin read *Natural Theology* while at Cambridge and was encouraged by his instructors John Henslow and Adam Sedgwick to accept Paley's perspective. Darwin recalled that Paley's work, including *Natural Theology*, "was the only part of the academical course which, as I then felt and as I still believe, was of the least use to me in the education of my mind." When Darwin boarded the *Beagle*, he accepted design in nature. However, after discovering natural selection, he felt differently: "The old argument of design in nature, as given by Paley, which formerly seemed to me so conclusive, fails, now that the law of natural selection has been discovered."

Virtually all biologists have similarly rejected Paley's argument. The most famous of these refutations is Richard Dawkins' *The Blind Watchmaker* (1986), whose title refers to Paley's metaphor. Dawkins agrees that there is a watchmaker but otherwise concludes that Paley is "gloriously and utterly wrong." The watchmaker for Dawkins (and for contemporary biology) is natural selection. Biologists view the evolution of complexity and apparent design, therefore, simply as the result of the cumulative process of repeated generations of differential reproduction. Dawkins' book motivated Phillip Johnson to write *Darwin on Trial* and to become active in the intelligent design (ID) movement. Although proponents of ID claim that their premises differ from Paley's, and that they, unlike Paley, do not specify who or what the designer is, most evolutionary biologists see ID as a version of Paley's arguments updated to account for advances in our understanding of biology.

Paley suffered for many years from a serious intestinal ailment. Soon after finishing *Natural Theology*, he suspected that his death was imminent, and he assembled his sermons to be published posthumously and given to anyone "likely to read them." Paley's suspicions of his impending death were borne out, and he died on May 25, 1805. He was buried in the Carlisle Cathedral, next to his wife.

STEPHEN MEYER (b. 1958)

Stephen C. Meyer is Vice President of the Discovery Institute and director of the Institute's Center for Science and Culture (CSC). He earned undergraduate degrees in geology and physics from Whitworth College and a Ph.D. in the history and philosophy of science from Cambridge University. While at Cambridge, Meyer met Phillip Johnson, who was then working on the manuscript that eventually became *Darwin on Trial* (1991). When the two men had returned to the United States, they joined with others to coalesce the intelligent design (ID) movement in the 1980s. Meyer and political scientist John West (now associate director of the CSC) were asked by Discovery Institute founder Bruce Chapman to establish a research center for ID. This was the start of the Center for the Renewal of Science and Culture, later renamed the Center for Science and Culture. The CSC "supports research by scientists and other scholars challenging various aspects of neo-Darwinian theory."

In 2004, Meyer published an article in the *Proceedings of the Biological Society of Washington* that reviewed preexisting information on the Cambrian explosion as evidence for ID. The manuscript did not follow the journal's standard peer-review process, being reviewed and accepted solely by the journal's editor, Richard Sternberg. Sternberg is a proponent of ID and a signatory to the Discovery Institute's "A Scientific Dissent from Darwinism." When they reviewed Meyer's article, the publishers of the journal deemed the article "inappropriate" for its publication. The Discovery Institute cites this response as evidence for "what amounts to a doctrinal statement in an effort to stifle scientific debate" by the scientific community but also dismisses the importance of peer review (i.e., "as if such journals represented the only avenue of legitimate scientific publication"). Meyer coedited (with John Angus Campbell) *Darwinism, Design and Public Education* (2004), which "presents a multi-faceted scientific case for the theory of intelligent design." Meyer is also a chief architect of the "teach the controversy" initiative of the Discovery Institute.

STATE OF TENNESSEE v. JOHN THOMAS SCOPES

State of Tennessee v. John Thomas Scopes (1925), the original "Trial of the Century," resulted in coach and substitute science teacher John Scopes being convicted of the misdemeanor of teaching human evolution in a public school in Tennessee. Scopes's trial, which William Jennings Bryan described as "a duel to the death" between evolution and Christianity, remains the most famous event in the history of the evolution-creationism controversy. The Scopes "monkey" trial also provided a framework for the play and movie *Inherit the Wind*.

JOHN THOMAS SCOPES v. STATE OF TENNESSEE

John Thomas Scopes v. State of Tennessee (1927) ended the legal issues associated with the Scopes trial when the Tennessee Supreme Court upheld the constitutionality of a Tennessee law forbidding the teaching of human evolution but urged that Scopes's conviction be set aside. Laws banning the teaching of human evolution in Tennessee, Mississippi, and Arkansas remained unchallenged for more than 40 years.

JOHN THOMAS SCOPES (1900–1970)

The basic freedoms of speech, religion, academic freedom to teach and to think for oneself defended at Dayton are not so distantly removed; each generation, each person must defend these freedoms or risk losing them forever.

John Thomas Scopes was born in Paducah, Kentucky, on August 3, 1900. Scopes's socialist father Thomas, whom Scopes described as "a very religious man," was especially influential in his development. Scopes graduated from high school in 1919 in Salem, Illinois; William Jennings Bryan, whose hometown was Salem, spoke at Scopes's high school commencement (Scopes remembered Bryan as "one of the most perfect speakers I have heard"). Scopes completed his first year of college at the University of Illinois, but health problems forced him to transfer to the University of Kentucky in 1920. After attending college there sporadically, Scopes received his A.B. in Arts-Law on June 2, 1924. Just a few days before schools opened in the fall of 1924, the coach and algebra teacher in Dayton, Tennessee, resigned, and Scopes—the first qualified applicant—was offered, and eagerly accepted, the job.

In April 1925, Scopes was asked to substitute teach for W. F. Ferguson, the high school's biology teacher. Scopes taught students about the various kingdoms of life, and on April 23 he told students to read a chapter in *A Civic Biology* (the course textbook) about evolution. However, Scopes was sick the next day, and never taught the subject.

Two weeks later, Scopes was playing tennis when he was summoned by local engineer George Rappleyea to Robinson's Drug Store to meet with a few civic leaders, including Walter White, Wallace Haggard, and Sue Hicks. Scopes admitted that he included evolution in his biology class, after which he was asked if he would be willing to be arrested to test the newly passed "Butler Law" banning the teaching of human evolution in Tennessee's public schools. When told that a trial could help the Dayton area, Scopes agreed to be charged with the crime and went back to his tennis game. On May 7, Scopes was charged with violating the Butler Law. When Bryan volunteered to help prosecute Scopes, famed defense attorney Clarence Darrow volunteered to help defend Scopes, and the stage was set for the most famous event in the history of the evolution-creationism controversy.

The jury ruled in *State of Tennessee v. John Thomas Scopes* in 1925 that Scopes had indeed violated state law. His conviction, however, was later overturned on a technicality by the Tennessee State Supreme Court in 1927 in *John Thomas Scopes v. State of Tennessee*.

Scopes had planned to leave Dayton for the summer and sell cars but was available for his trial because he decided to stay in town to date "a beautiful blond" he had met at a church function. Scopes participated in several pretrial publicity stunts (for example in New York and Philadelphia), and his trial became a sensational event worldwide. During the trial, which pitted science against religion, Scopes took the issues seriously, but he was largely irrelevant; on some days the proceedings began without him. Scopes, who described the trial as "just a drugstore discussion that got past control," did not testify at his trial, and was eventually convicted. He was fined $100, which the Baltimore *Sun* (which had sent reporter H. L. Mencken to cover the trial) agreed to pay. The "Bill of Cost" for the Scopes trial totaled $343.87, of which $16.10 was charged to Rhea County. Scopes was billed for the rest, but he never paid. In his only statement to the court, Scopes said, "Your Honor, I feel that I have been convicted of violating an unjust statute. I will continue in the future, as I have in the past, to oppose this law in any way I can. Any other action would be in violation of my idea of academic freedom. . . . I believe the fine is unjust."

Scopes left Dayton a few days after his famous trial and was at a train station in Knoxville when he learned of Bryan's death. Scopes returned to Dayton to view the bronze casket containing Bryan's body as it lay in state in the F. R. "Richard" Rogers home.

Scopes received thousands of letters, including proposals for marriage, an offer to become a minister of a new church and "Bishop of Tennessee" with "pontifical powers," lucrative offers to lecture and star in movies, and advice for salvation. Scopes burned the letters (most of them unopened) and refused to cash in on his accidental fame, noting that he simply wanted "peace and emotional stability." Scopes was content to let his hour of fame pass; as he noted later in his life, "I had only one life in this world and I wanted to enjoy it. I knew I could not live happily in a spotlight. . . . [I wanted to be] just another man instead of the Monkey Trial defendant."

After the trial, F. E. Robinson offered Scopes a new contract to teach at Rhea County High School (at a salary of $150 per month), provided he would "adhere to the spirit of the evolution law." Scopes declined the offer and never taught again. He was replaced at Rhea County High School by Raleigh Reece, a reporter from Nashville who had asked prosecutor Sue Hicks for some "inside dope" while covering Scopes's trial for the *Nashville Tennessean*. Scopes accepted a scholarship (funded by scientists and newsmen who attended his trial) to study geology at the University of Chicago in September, 1925. While living in Chicago, Scopes became close friends with Clarence Darrow, noting later that except for his father, "Darrow had a greater influence on my life than any other man I have known."

Scopes hoped to earn a Ph.D., and in 1927 applied for a fellowship to fund his work (the scholarship money raised by the scientists and reporters from Dayton lasted only two years). However, the president of the University of Chicago responded to Scopes's application with a terse letter: "Your name has been removed from consideration for the fellowship. As far as I am concerned, you can take your atheistic marbles and play elsewhere." Realizing that he had to abandon his goal of teaching and turn to commercial work, Scopes was hired by Gulf Oil and sent to do fieldwork in northwest Venezuela near the city of Maracaibo. There, at a dance, Scopes met Mildred Walker, "a pretty brown-eyed brunette from South Carolina" (and fellow employee of Gulf Oil). Walker, whose aunt "thought that Scopes was something with horns," married Scopes in a Catholic service in February 1930, in Maracaibo. In 1933, Scopes, an agnostic, returned to Texas (and later to Louisiana) to work for Union Producing Company, a subsidiary of United Gas Corporation. Throughout his career, Scopes's modest office included no mementos of his famous trial. John and Mildred had two sons, William and John Jr., both of whom worked in the insurance industry. Their father spoke with them about his famous trial only when asked.

When the movie version of *Inherit the Wind* opened in 1960, Scopes returned to Dayton, where he bought a "Scopes soda" that was "priced now as then in honor of Scopes's return to Dayton." Although a local preacher denounced Scopes as "the devil," other festivities—including a parade, concert, and car show—were more cordial. Scopes listened as Mayor J. J. Rogers—in front of the second-largest crowd in

Dayton's history—gave him a key to the city and proclaimed the day Scopes Trial Day.

Scopes retired in 1963 and four years later published his memoirs, *Center of the Storm: Memoirs of John T. Scopes*. In 1968, Scopes promoted his book with interviews and television appearances (e.g., *The Today Show, The Tonight Show with Johnny Carson*), during which Scopes noted that "the Bible simply isn't a textbook of science, and that's all there is to it." When asked about *Inherit the Wind*, Scopes noted that the controversy would "go on, with other actors and other plays."

In 1969, Scopes met in Shreveport, Louisiana, with Susan Epperson, whose challenge to the Arkansas antievolution law (*Epperson v. Arkansas*) had resulted in the U.S. Supreme Court ruling unanimously that laws banning the teaching of human evolution in public schools are unconstitutional. Epperson believed that the Arkansas ban on the teaching of evolution was "a sure path to the perpetuation of ignorance, prejudice, and bigotry," and her principal brief to the Court closed with a dramatic reference to "the famous Scopes case" and to the "darkness in that jurisdiction" that followed Scopes's verdict. When Scopes heard of the Court's decision, he claimed that "this is what I've been working for all along. . . . I'm very happy about the decision. I thought all along—ever since 1925—that the law was unconstitutional."

On April 1, 1970, Scopes returned to a Tennessee classroom for the first time in 45 years when he accepted an invitation to talk to biology students at Vanderbilt University's Peabody College for Teachers. Scopes, who was greeted "like a returning hero," reiterated his belief that "it is the teacher's business to decide what to teach. It is not the business of the federal courts nor of the state." Forty-five years after his trial, Scopes received an average of one letter per day about his famous trial.

Scopes's talk at Vanderbilt was his last public appearance. In July 1970, Scopes—a heavy smoker—was diagnosed with inoperable cancer. He underwent radiation treatments but died at his home in Shreveport, Louisiana, on October 21, 1970. Following a service at Shreveport's St. John's Catholic Church, Scopes's remains were returned to Paducah, Kentucky, and buried beneath the inscription "A Man of Courage," a phrase used to describe Scopes by Clarence Darrow. Scopes rests in a family plot in Oak Grove Cemetery, not far from where he first learned biology in elementary school. He rests beside his parents (Thomas [1860–1945] and Mary Brown Scopes [1865–1957]), wife Mildred (1905–1990), and sister Lela (1896–1989), who lost her job as a math teacher in Paducah when she would not renounce her brother's beliefs. In 1990, Scopes's wife Mildred—also a heavy smoker—died of emphysema in the same room in which her husband had died 20 years earlier.

EUGENIE SCOTT (b. 1945)

Eugenie Scott was born on October 24, 1945, in Lacrosse, Wisconsin. When she was 10, Scott—a compulsive reader—picked up her older sister's college anthropology textbook. She was fascinated with what she learned and decided to become an anthropologist.

After earning a B.S. (1967) and an M.S. (1968) at the University of Wisconsin-Milwaukee, Scott enrolled as a doctoral student in anthropology at the University of Missouri-Columbia. Scott completed her Ph.D. in 1974, after which she joined the faculty in the Department of Anthropology at the University of Kentucky. While there, Scott took some of her students to attend a debate between Jim Gavan (a former president of the American Association of Physical Anthropology) and antievolutionist Duane Gish of the Institute for Creation Research. That debate, which Scott described as "an eye-opener," showed Scott that creationists were intent on undermining her profession and the scientific point of view.

In 1980, Scott experienced what she called her "true baptism" regarding the social and political importance of the evolution-creationism controversy when a group calling itself "Citizens for Balanced Teaching of Origins" asked the Lexington (Kentucky) Board of Education to include "creation science" in its curriculum. Scott's efforts helped convince the Board to reject the creationists' request (by a 3–2 vote) and taught Scott that creation science is not a problem that can be solved with scientific evidence alone.

In 1986, as the "creation science" movement raged in the United States, Scott became director of the National Center for Science Education, Inc. (NCSE), a small, not-for-profit organization that provides information, advice, and resources for schools, parents, educators, and concerned citizens dedicated to keeping evolution in the science curriculum of public schools. Under Scott's leadership, the NCSE soon became the premier organization dedicated to supporting the teaching of evolution. Scott's work has been recognized with awards from organizations such as the National Association of Biology Teachers and the American Association for the Advancement of Science (AAAS). Scott has authored *Evolution v. Creationism: An Introduction* and is coeditor of *Not in Our Classrooms: Why Intelligent*

Design Is Wrong for Our Schools. Today, Scott—who views intelligent design as a "soft-core antievolution strategy"—is a self-described "evolution evangelist."

PHILLIP JOHNSON (b. 1940)

This isn't really, and never has been a debate about science. It's about religion and philosophy.

Phillip E. Johnson was born in 1940 in Aurora, Illinois. His family was only nominally religious, and in college he was agnostic. Johnson entered Harvard University when he was 16 years old and graduated with a degree in English in 1961. After graduation, Johnson taught English in East Africa.

When he returned home, Johnson found that his father had submitted an application for him to the University of Chicago Law School. He was accepted and graduated in 1965 at the head of his class. He clerked for Chief Justice Robert Traynor of the California Supreme Court and then for Chief Justice Earl Warren of the United States Supreme Court. Johnson accepted a faculty position at the Boalt School of Law at the University of California-Berkeley in 1967 and remained at Berkeley for the rest of his career (although he also served as Deputy District Attorney for Ventura County while on leave from teaching). He retired in 2000 and currently is Jefferson E. Peyser Professor of Law Emeritus at Berkeley. In the mid-1980s, Johnson was considered to fill a vacancy on the Ninth Circuit Court of Appeals.

Although he had achieved professional success, in the 1970s Johnson perceived a lack of meaning in his life. His marriage dissolved around this time and he was unfulfilled by what he labeled "superficial" interactions he had with colleagues. While visiting his 11-year-old daughter at a Bible camp, Johnson became an evangelical Christian. It was through his new church that he met his second wife.

While on sabbatical in 1987 at University College, London, Johnson read Richard Dawkins' *The Blind Watchmaker*. Johnson admired Dawkins' rhetorical abilities but found Dawkins' argument for the power of unguided evolution unconvincing. As Johnson saw it, evolutionary biologists did not consider all the evidence (i.e., especially that for design) and instead were confusing assumptions and conclusions. Johnson also claimed that scientists accept evolution only because they have a dogmatic, *a priori* commitment to naturalism. According to Johnson, science has therefore not shown that God does not exist; it merely assumes there is no God because to do otherwise is not "science." Accepting evolution then necessarily means adhering to an atheistic worldview, because one cannot simultaneously accept evolution and believe in God.

Johnson began writing a manuscript that considered evolution from a lawyer's perspective: What does the evidence indicate as to the likelihood of this concept explaining life as we know it? *Darwin on Trial* (1991) concluded that evolution fails to explain the observed patterns we see in nature. Not surprisingly, the book was rejected by scientists. The review by Harvard paleontologist Stephen Jay Gould in *Scientific American* was typical; it called the book "full of errors, badly argued, based on false criteria, and abysmally written." (Johnson submitted a rebuttal to Gould's review, but *Scientific American* declined to publish it.) Scientists noted that acceptance of evolution does not endorse atheism (e.g., Gould's well-known "non-overlapping magesteria") and claimed that Johnson was confusing the methodological materialism required of science with the philosophical materialism that some scientists (including Dawkins) propose flows from scientific understanding.

Johnson attacked naturalism as science's "creation myth" in his next book, *Reason in the Balance* (1995), and introduced the mutually exclusive options of "methodological naturalism" (i.e., contemporary science) and "theistic realism." Johnson's theistic realism assumed that "the universe and all its creatures were brought into existence for a purpose by God." This perspective is not scientific, nor does it accommodate belief systems that reconcile religion and evolution, such as theistic naturalism, which claims that God has created and works through natural forces, including evolution.

Johnson also began developing a strategy to undermine evolution. In 1992, members of the nascent intelligent design (ID) movement met at Southern Methodist University (SMU). The meeting included Johnson, Michael Behe, Stephen Meyer, and William Dembski, all of whom became leading proponents of ID. The meeting represented a pivotal moment in the development of ID because it was there that the "wedge strategy" was first proposed.

The wedge strategy ("a log is a seeming solid object, but a wedge can eventually split it by penetrating a crack and gradually widening the split. In this case the ideology of scientific materialism is the apparently solid log"), honed over the next few years at meetings similar to the one at SMU (but kept secret from public scrutiny), was well developed by the mid-1990s. A home for the strategy was created in 1996 when the Discovery Institute, a Seattle-based Christian think tank founded by Bruce Chapman (former Secretary of State for the state of Washington), announced the establishment of the Center for the Renewal of Science and Culture, later renamed the Center for Science and Culture (CSC). Johnson served as Program Advisor and Stephen Meyer as Director, with most other notable ID proponents as Fellows of the Center. Johnson was the leading edge of the wedge, or in his words, "I'm like an offensive lineman in pro football. . . . [T]he idea is to open up a hole that a running back can go through." Although Johnson has at times emphasized keeping the discussion purposely vague about who or what the "designer" is (even claiming that it could be an alien that brought life to Earth), at other times he has claimed that he wants to "get the issue of intelligent design, which really means the reality of God, before the academic world and into the schools."

The wedge strategy was encapsulated in the "wedge document" (whose existence the CSC initially denied), which outlined five-year and twenty-year goals for the "overthrow of materialism and its cultural legacies." The plan included three phases: scientific research, writing, and publication; publicity and opinion-making; and cultural confrontation and renewal. Despite claims to the contrary by the CSC, no peer-reviewed research relative to ID has appeared in the scientific literature. However, an article by CSC director Stephen Meyer was published in the *Proceedings of the Biological Society of Washington* in 2004. This article reviewed existing information—no new scientific data were presented—and the managing editor alone decided to publish it (i.e., there was no peer review). The publishers of the journal issued a statement stating that, given the opportunity, they "would have deemed the paper inappropriate" for publication.

Johnson and his colleagues have implemented the other two phases of the wedge strategy, publicity and public confrontation. Johnson has written seven books discussing evolution and, until recently (he suffered a series of strokes in 2002 and 2005), traveled frequently to speak about ID and the struggle against scientific naturalism. In 1999, in the wake of the Kansas State Board of Education's decision to abandon the teaching of evolution, Johnson helped craft a modified tactic— "teach the controversy." This strategy shifted the focus from teaching ID to portraying evolution as a controversial theory in trouble; any discussion of evolution therefore should include its (allegedly) serious problems and an examination of alternatives. "Teach the controversy" is often included in legislation to promote ID in public school science curricula.

Johnson also authored the "Santorum Amendment" that former U.S. Senator Rick Santorum of Pennsylvania tried to attach to the 2001 Elementary and Secondary Education Act Authorization Bill (later renamed the No Child Left Behind Act). The amendment invoked standard "teach the controversy" ideals about science education and specified evolution as a topic needing such discussion. Although the Discovery Institute supported the amendment, it was not included in the final version of the bill enacted into law.

MICHAEL BEHE (b. 1952)

If the evidence was there, people would be quieter, but the evidence isn't there, and people know it.

Michael J. Behe was born on January 18, 1952, in Altoona, Pennsylvania. Behe was taught evolution in the Catholic schools he attended, and as a young man accepted a universe that was created by a god working through observable natural laws, including natural selection: "Here was Darwin's theory, and it looks like God set up the world to begin producing life. I remember thinking 'That's cool.' "

Behe earned a B.S. from Drexel University in chemistry in 1974 and a Ph.D. in biochemistry from the University of Pennsylvania in 1978. Following postdoctoral work at the National Institutes of Health, Behe became an assistant professor in the chemistry department at Queens College (part of the City University of New York) in 1982. There he met his future wife, Celeste, with whom he eventually had nine children. In 1985, Behe became an associate professor of chemistry at Lehigh University in Bethlehem, Pennsylvania. In 1995, he moved to the biology department at Lehigh, and became full professor in 1997.

By the 1990s, Behe had published several dozen articles in peer-reviewed journals. During this phase of his career, Behe accepted the central role of evolution in producing the processes he studied. However, in 1987 he read Michael Denton's *Evolution: A Theory in Crisis* (1985). Denton, a biochemist, concluded that there is no mechanism that would cause speciation, and rejected Darwin's proposal of common descent. Behe's response to Denton's book was that it "startled me. . . . [T]here was a good chance [Darwin's theory] was incorrect; it could not really describe how life came to be. I got mad." Behe searched for research that demonstrated how complex biochemical systems could have evolved. He concluded that such support was lacking, which led him to believe that complex biological systems must have been designed by an external agent.

In the early 1990s, Behe read Phillip Johnson's *Darwin on Trial* (1991), finding in it much that agreed with his new perspective on evolution. After a letter by Behe in support of Johnson's book was published in *Science,* Johnson invited Behe to join others in advocating intelligent design (ID). Behe accepted and contributed material on biochemistry to the 1993 version of the pro-ID textbook *Of Pandas and People: The*

Central Question of Biological Origins. In 1996, Behe also became a senior fellow at the pro-ID Discovery Institute.

In 1996, Behe described his opposition to standard evolutionary biology in his iconoclastic *Darwin's Black Box: The Biochemical Challenge to Evolution.* This book sold more than 200,000 copies, was named Book of the Year for 1996 by *Christianity Today,* and was hailed by the *National Review* as one of the top 100 nonfiction books of the twentieth century. The "black box" of the title is the basic unit of all life, the cell. Behe claims that when Darwin proposed his thesis, the cell was relatively unknown and was considered a simple building block upon which the complexity of life was constructed. However, contemporary research has shown that cells are complicated, meaning that life is not "complexity at the top and simplicity beneath, but . . . complexity at the top and more complexity underneath." Behe proposed that it is impossible to imagine how the intricate "molecular machines" of the cell could have evolved by natural selection because only the final version with all the necessary parts in place produces a working system: cellular components (like Behe's favorite example, the flagella) are therefore "irreducibly complex" and are scientific evidence for the action of an intelligent designer. *Darwin's Black Box* soon became a foundation for ID's "scientific" opposition to evolution.

Response to *Darwin's Black Box* from the scientific community was swift and severe. Behe was criticized for not understanding basic evolutionary biology (e.g., he has described the action of natural selection as "random Darwinian processes" and has noted that "cells are simply too complex to have evolved randomly"), as well as for mixing science and religion by invoking a supernatural cause that he claims can be studied scientifically. His concept of irreducible complexity has been equated with both the discredited "argument from design" proposal (i.e., traits appear to be designed because they *are* designed) and the weak "God in the gaps" argument for the existence of God (i.e., the inability of science to explain all aspects of the natural world is proof of God). Behe dismissed his critics' evolutionary explanations as "materialism in the gaps." Although Behe claims that ID is "one of the greatest achievements in science," most scientists regard it as religion, not science.

Behe is unusual among ID proponents in that he accepts an evolutionary explanation for many traits, agrees that the Earth is ancient, and recognizes common descent among organisms, including between humans and apes. But critics charge that he inexplicably treats evolution at the cellular level differently (e.g., "Darwin's theory encounters its greatest difficulties when it comes to explaining the development of the cell") and promotes the need for entirely different explanations at these levels. Although Behe claims otherwise, the scientific community has deemed ID and irreducible complexity as not scientific because the concepts do not generate testable hypotheses. The lack of publication of original, peer-reviewed research in this area is also cited as evidence for the nonscientific nature of ID.

Behe counters that he has tried to publish ID-based research and to obtain grant money for study of ID, but his efforts are consistently rejected. In 2004, Behe and David Snoke, a professor in the department of Physics and Astronomy at the University of Pittsburgh, did publish a paper in the peer-reviewed journal *Protein Science* titled "Simulating Evolution by Gene Duplication of Protein Features That Require Multiple Amino Acid Residues." In their introduction, Behe and Snoke make clear the purpose of their article: "Although many scientists assume that Darwinian processes account for the evolution of complex biochemical systems, we are skeptical." The authors do not mention "intelligent design" or "irreducible complexity" but do report that for mutation to produce novel protein functions, large populations would be required. They conclude that "such numbers seem prohibitive" but urge caution in drawing conclusions from this single study. A critical response in the same journal by evolutionary biologist Michael Lynch of Indiana University deemed the model used by Behe and Snoke as non-Darwinian because it assumed intermediate steps between the original and new protein to be non-functional. Lynch created a "Darwinian version" of the model and concluded that "the origins of new protein functions are easily explained in terms of well-understood population-genetic mechanisms."

Behe was queried extensively about the *Protein Science* paper during his multiday testimony in the *Kitzmiller v. Dover Area School District* trial in 2005. Behe testified that he considered the article to be an "intelligent design article" that "seems to present . . . problems for Darwinian evolution," and that the original version did in fact include the term "irreducible complexity," but a reviewer required the term be removed. When cross-examined by the plaintiffs' attorney, Behe confirmed that the model was limited to only one particular type of mutation and that the model did indeed predict that advantageous mutations could become fixed in the population in a reasonable number of generations within typically large microbial populations. Other aspects of Behe's testimony figured prominently in Judge John Jones' final decision:

> Consider, to illustrate, that Professor Behe remarkably and unmistakably claims that the plausibility of the argument for ID depends upon the

extent to which one believes in the existence of God. As no evidence in the record indicates that any other scientific proposition's validity rests on belief in God, nor is the Court aware of any such scientific propositions, Professor Behe's assertion constitutes substantial evidence that in his view, as is commensurate with other prominent ID leaders, ID is a religious and not a scientific proposition.

Behe's viewpoints remain controversial; even his home department's Web site includes a disclaimer stating that "intelligent design has no basis in science, has not been tested experimentally, and should not be regarded as scientific." Behe, along with fellow creationists William Dembski and Stephen Meyer, edited *Science and Evidence for Design in the Universe* (2001), a collection of essays addressing the role of design in nature. In 2007, Behe published *The Edge of Evolution: The Search for the Limits of Darwinism*, a follow-up to *Darwin's Black Box*, which propounds Behe's views on the limited creative power of mutation and natural selection. Like *Darwin's Black Box*, *The Edge of Evolution* was harshly criticized by scientists. Behe continues to write about ID for the popular press and has produced several videos about ID that are distributed through the Access Research Network.

NATIONAL CENTER FOR SCIENCE EDUCATION (Est. 1981)

The National Center for Science Education (NCSE) is a not-for-profit, membership-based organization that provides information, advice, and resources for schools, parents, educators, and concerned citizens dedicated to keeping evolution in the science curriculum of public schools. The NCSE is religiously neutral and works with local and national religious, educational, and scientific organizations such as the Center for Theology and the Natural Sciences, the National Association of Biology Teachers, and the National Academies of Science.

The NCSE's origins can be traced to the late 1970s and early 1980s, when many citizens were appalled to learn that their state legislatures were considering legislation mandating "balanced treatment" for "creation science." In 1980, Iowa high school teacher Stanley Weinberg (1911–2001) began organizing statewide Committees of Correspondence "committed to the defense of education in evolutionary theory." Like their namesakes in the Colonial era, these Committees shared information about policy-related issues—in this instance, those underlying legislative attempts to undermine the teaching of evolution. In 1981, individuals who had been involved in several of these Committees founded the NCSE, and in 1983 the NCSE was incorporated with Weinberg as its first president. Today, the NCSE has more than 4,000 members and an annual budget of $700,000.

The NCSE, which is based in Oakland, California, is involved in numerous activities to promote the teaching of evolution in public schools. The organization also publishes the bimonthly journal *Reports of the National Center for Science Education.*

NATIONAL ASSOCIATION OF BIOLOGY TEACHERS (Est. 1938)

The National Association of Biology Teachers (NABT) is the largest national association dedicated to the interests of biology teachers. The NABT was conceived by Oscar Riddle of the Carnegie Institute, which provided $10,000 for the new organization. The first edition of the NABT's journal, *The American Biology Teacher*, appeared in October 1938. Today, the NABT has more than 8,000 members in the United States and abroad. Since its inception, the NABT has been active in the evolution-creationism controversy. The organization supported the lawsuits of Susan Epperson and Don Aguillard, and has opposed attempts to legislate creationism and undermine evolution (e.g., *McLean v. Arkansas Board of Education*).

Late in 1970, *The American Biology Teacher* published an article by creationist Duane Gish pleading for teachers to teach creationism. When the journal published another article by Gish three years later, the article was prefaced with a statement noting that biologists reject creationism (this article appeared in the same volume as the famous article by Theodosius Dobzhansky titled "Nothing in Biology Makes Sense Except in the Light of Evolution").

In 1972, the NABT helped form state-based "Committees of Correspondence on Evolution" to resist creationists' efforts to undermine science. In 1981, several of these Committees of Correspondence founded the National Center for Science Education. Near the same time, the NABT's "Fund for Freedom in Science Teaching" gathered more than $12,000 to oppose the teaching of creationism, but there was also a backlash from NABT members who were creationists. In response to these complaints, the organization sponsored a creationism panel at its annual meeting the following year that included presentations by creationists such as Duane Gish.

In 1995, the NABT adopted a statement about evolution whose preamble described evolution as "an unsupervised, impersonal, unpredictable and natural process." When some scientists complained that the words *unsupervised* and *impersonal* implied theology, the offending words were deleted from the statement. Today, the "NABT Statement on Teaching Evolution" endorses the teaching of evolution and opposes the teaching of creationism.

NATIONAL ACADEMY OF SCIENCES (Est. 1863)

There is no significant scientific doubt about the close evolutionary relation-ships among all primates or between apes and humans.

The National Academy of Sciences (NAS), which originated during President Abraham Lincoln's administration, is a self-selected, honor-ific society of approximately 1,800 members and 350 foreign associates who advise national leaders about scientific issues. In 1916, at the request of President Woodrow Wilson, the Academy established the National Research Council to provide technological recommendations regarding military preparedness. The NAS advocates evolution educa-tion from kindergarten through college and has published a variety of evolution-related books, including *Teaching about Evolution and the Nature of Science* (1998), *The Future of Evolution* (2002), *Science and Creationism: A View from the National Academy of Sciences* (second edi-tion, 1999), and *Tempo and Mode in Evolution: Genetics and Paleontology 50 Years After Simpson* (1995). Although fewer than 10 percent of NAS members believe in a personal God, the NAS notes that many scientists believe in God and accept evolution.

NATIONAL SCIENCE FOUNDATION (Est. 1950)

The National Science Foundation (NSF) is a federal agency created to promote science for the improvement of life. The NSF supports basic research and science education, and has funded many projects that have promoted evolution. NSF's publication in 1970 of the pro-evolution curriculum *Man: A Course of Study (MACOS)* produced a national uproar; Congressman John Conlan and Senator Jesse Helms condemned the project and urged the NSF to cut funding for evolution-related projects. In 1973, Congress considered—but did not pass—legis-lation giving it direct supervision and veto power over every project funded by the NSF. In 1980, Ronald Reagan cited *MACOS* as evidence that the government was endorsing subversive values, and urged the NSF to develop Christian curriculum materials. In 1989, the NSF asked Biological Sciences Curriculum Study director Joe McInerney to remove the word *evolution* from the title of a project because some members of Congress would be unhappy to learn that the NSF had funded an evolution-related project. McInerney ignored the NSF's request.

NATIONAL SCIENCE TEACHERS ASSOCIATION (Est. 1944)

The National Science Teachers Association (NSTA) originated as a department of the National Education Association, which endorsed the teaching of evolution in 1916. In the 1960s, the NSTA supported

Susan Epperson's attempt to overturn the Arkansas law banning the teaching of human evolution, noting that evolution "is firmly established even as the rotundity of the Earth is firmly established." In subsequent years, the NSTA continued to support the teaching of evolution, and supported Gary Scott when he was fired for teaching evolution in Tennessee in 1967. NSTA headquarters are located in Arlington, Virginia.

JEAN-BAPTISTE LAMARCK (1744–1829)

Do we not therefore perceive that by the action of the laws of organization . . . nature has in favorable times, places, and climates multiplied her first germs of animality, given place to developments of their organizations and increased and diversified their organs? Then . . . aided by much time and by a slow but constant diversity of circumstances, she has gradually brought about in this respect the state of things which we now observe. How grand is this consideration, and especially how remote is it from all that is generally thought on this subject?

Jean-Baptiste Pierre Antoine de Monet, Chevalier de Lamarck was born into a military family on August 1, 1744, at Bazentin-le-Petit, Picardy, in rural northern France. He was the youngest of eleven children. Phillipe, Lamarck's father, expected Jean to have a church-related career, and in 1756 Lamarck enrolled in a Jesuit seminary. However, when Phillipe died in 1760, Lamarck quit the seminary, bought a horse, and joined the French army. In his first battle, Lamarck distinguished himself for bravery and was made an officer. However, an injury inflicted by a comrade forced Lamarck out of the military, after which he worked as a bank clerk in Paris. Lamarck then began to study botany and medicine.

At age 34, Lamarck published *Flore Francaise*, an acclaimed book about the plants of France. The following year—with the help of Georges-Louis Buffon, whose son Lamarck tutored—Lamarck was elected to the French Academy of Science. In 1793 (the same year that Louis XVI and Marie Antoinette went to the guillotine), Lamarck helped reorganize the French Museum of Natural History, and was appointed a professor there the following year. The museum was to be run by 12 professors in 12 scientific fields, and Lamarck was in charge of studying "insects, worms, and microscopic animals." Lamarck knew little about insects and worms, but he later coined the word *invertebrate* to describe them. Today, a plaque at the museum's entrance notes that Lamarck lived there from 1795 until his death in 1829.

Lamarck first presented his ideas about evolution in a lecture on May 11, 1800. In 1809, Lamarck—who was a protégé of Buffon and botanist to King Louis XVI—published his ideas about evolution in his most famous work, *Philosophie Zoologique*. In this purely theoretical book, which Darwin read while aboard the *Beagle*, Lamarck discussed life's "tendency to progression" and "tendency to perfection" and claimed that life is in a constant state of advancement and improvement that is too slow to be perceived except with the fossil record. Lamarck's claim required spontaneous generation of new species to replace those transformed to more advanced species. Whereas contemporaries such as Buffon had hinted at evolution, Lamarck was its champion: ". . . [S]pecies have only a limited or temporary constancy in their characters . . . there is no species which is absolutely constant." Lamarck was confident of his conclusions; as he once noted, "I am not submitting an opinion, but announcing a fact."

Lamarck, who coined the term *biology*, was perhaps the world's premier invertebrate zoologist. He argued that organisms contained a "nervous fluid" that enabled them to adapt to their local environments. Lamarck believed that organisms evolve to become more complex over time; that these purposeful changes are brought about by the use or disuse of acquired traits; and that these changes made the organisms better able to survive in new environments and conditions. Simply put, Lamarck believed that what an animal did during its lifetime was passed on to its offspring.

According to Lamarck, environmental changes alter the needs of organisms living in that environment. In turn, the organisms' altered needs change the organisms' behaviors, and these altered behaviors then lead to the greater or lesser use of different structures. The more an organism used a part of its body, the more developed that part would become (similarly, the disuse of a part would result in its decay). Lamarck referred to this idea—namely, that the use or disuse of a structure would cause the structure to develop or shrink—as his "First Law." This was followed by Lamarck's "Second Law," which proposed that the changes acquired as a result of the First Law would be inherited by the organism's offspring. As a result, species would gradually change as they became adapted to their environment. For example, Lamarck argued that wading birds evolved long legs as they stretched them to keep high and dry. Similarly, giraffes evolved long necks as they stretched their necks to reach leaves high in trees. When they stretched, Lamarck claimed, their "nervous fluid" would flow into their necks and, over successive generations, cause their necks to grow longer. This inevitable need-based change (i.e., necks getting longer to get food) would give giraffes permanently longer necks, and these long necks would be passed to the giraffes' offspring. Lamarck's idea, which came to be known as "inheritance of acquired characteristics," suggested that there was a drive toward perfection and complexity, analogous to species climbing a ladder. Lamarck rejected extinction, instead claiming that organisms evolved into different, more perfect species via a process he called "transmutation." New, primitive organisms constantly formed from inorganic matter at the bottom of the scale.

Lamarck—an ardent materialist—was his era's most renowned advocate of evolution, and his model for evolution was the first testable hypothesis to explain how a species could change over time. Many people—Erasmus Darwin among them—endorsed Lamarck's idea. However, it was rejected (and sometimes ridiculed) by the leading scientists of his time (e.g., his revered colleague Cuvier), and was later dismissed by other scientists. Nevertheless, Lamarck's idea was popular with the public—so much so that Charles Darwin alluded to it in later editions of his *Origin of Species*. Darwin wrote to Joseph Hooker in 1844 that "Heaven forfend me from Lamarck nonsense of a 'tendency to progression,' 'adaptations from the slow willing of animals,' etc.! But the conclusions I am led to are not widely different from his; though the means of change are wholly so. I think I have found out (here's presumption!) the simple way by which species become exquisitely adapted to various ends."

Although Lamarck's name is most often associated with his discredited "inheritance of acquired traits," Darwin and many other scientists acknowledged him as a great zoologist and one of evolution's early thinkers. Although Darwin described Lamarck's book as "veritable rubbish," Darwin conceded in 1861 that "Lamarck was the first man whose conclusions on the subject excited much attention. This justly celebrated naturalist . . . first did the eminent service of arousing attention to the probability of all changes in the organic, as well as in the inorganic world, being the result of law, and not of miraculous interposition." Lamarck's speculative suggestions regarding the origin of new traits continue to overshadow his otherwise important contributions to biology.

Lamarck married Marie Delaporte, the mother of his first six children, on her deathbed in 1792. In 1795 he then married Charlotte (who died in 1797), and in 1798 he married his third wife, Julie Mallet, who died in 1819. He is rumored to have married a fourth time, but no documents support this claim. When in his 70s, Lamarck's eyesight began to deteriorate. Despite his great contributions to biology, Lamarck spent his last decade a blind, penniless man living in obscurity and cared for by his unemployed daughters.

Lamarck died in Paris on December 18, 1829. His papers, books, and belongings were auctioned, and Lamarck was buried in a lime pit in Montparnasse cemetery with other paupers. Cuvier, who respected Lamarck's studies of invertebrates but rejected his theory of evolution, used his eulogy to ridicule and discredit Lamarck. Similarly, Lamarck's obituary in the London *Times* did not mention his many contributions to biology, instead focusing on the politics of finding his replacement at the museum. Although Lamarck's daughter claimed that "posterity will remember you," Lamarck's corpse was later excavated and piled with other nameless remains in the Paris catacombs. Today, Lamarck is memorialized with a large statue inscribed "Founder of the Doctrine of Evolution" at Jardin des Plantes in Paris. Far across the city is *rue Lamarck*, alongside which is the 86-meter *rue Darwin*.

RODNEY LeVAKE (b. 1954)

The process of evolution itself is not only impossible from a biochemical, anatomical, and physiological standpoint, but the theory of evolution has no evidence to show that it actually occurred.

Rodney LeVake was born on November 16, 1954, in Colorado Springs, Colorado. After attending St. John's University and double-majoring in Natural Science and Social Science, LeVake enrolled at Minnesota State University at Mankato, where he earned an M.A. in Teaching in Life Science. In 1984, LeVake began teaching math and science in Faribault Junior High School in Faribault, Minnesota, and in 1997, he began teaching biology at Faribault High School. The course syllabus, course registration guide, and the curriculum adopted by the school board listed evolution as part of the biology curriculum. When other science teachers suspected that LeVake had not taught evolution, LeVake was confronted by his friend and colleague Ken Hubert about his teaching of evolution. When LeVake responded on April 15, 1998, that he rejected evolution because he believed it was not scientific, he was reassigned to a ninth-grade physical science course.

LeVake, a self-described fundamentalist, felt that he had been treated unfairly. Seeking advice, LeVake contacted The Rutherford Institute, Focus on the Family, and the American Center for Law and Justice (ACLJ). The ACLJ, an organization founded by televangelist Pat Robertson to defend "the rights of believers," helped LeVake sue the school and its administrators, claiming that LeVake was reassigned because his religious beliefs opposed evolution. LeVake—a member of the Institute for Creation Research—asked the Court to give him

$50,000 (plus court costs) and declare "the district's policy, of excluding from biology teaching positions persons whose religious beliefs conflict with acceptance of evolution as an unquestionable fact, to be unconstitutional and illegal under the U.S. and Minnesota Constitutions." The District Court ruled against LeVake, noting that a teacher's right to free speech does not permit the teacher to circumvent the prescribed curriculum. LeVake began a lengthy appeal. The Minnesota Court of Appeals supported the original ruling, and LeVake's case ended on January 7, 2002, when the U.S. Supreme Court refused, without comment, to hear his case (*LeVake v. Independent School District #656*).

Today, LeVake still teaches at Faribault High School.

BIOLOGICAL SCIENCES CURRICULUM STUDY (BSCS) (Est. 1958)

The history of the Biological Sciences Curriculum Study (BSCS) can be traced to the October 4, 1957, launch of the small (23" diameter, with one watt of power) Soviet satellite *Sputnik*, which awakened the American political and educational establishment to the importance of improving science education. The following year, Congress passed the National Defense Education Act, which encouraged the National Science Foundation (NSF) to develop state-of-the-art science textbooks. In the same year, the NSF allocated $143,000 to establish the BSCS to educate "Americans in general to the acquisition of a scientific point of view." By 1959, BSCS had established its headquarters at the University of Colorado.

In the early 1960s, the BSCS created new biology textbooks that, unlike other textbooks, stressed concepts rather than facts and investigations rather than lectures. The three BSCS books published in 1963 became known by the color of their covers: *Blue* emphasized molecular biology, *Green* emphasized ecology, and *Yellow* emphasized cellular and developmental biology. Approximately 70 percent of the content of each book was identical, but the material was presented using different themes. Although the BSCS wanted to avoid the criticism that it was trying to establish a national curriculum, their books—for all practical purposes—did exactly that, for in the 1960s, most schools in the United States used BSCS textbooks.

When John Scopes was convicted of teaching human evolution in 1925, publishers feared that discussing evolution in biology textbooks would hurt sales. As a result, biology textbooks published after Scopes's conviction did not include the word *evolution*. However, BSCS books were different. Instead of relying on professional writers to prepare their textbooks, the BSCS recruited the best scientists and teachers in the United States as authors. Not surprisingly, all of the BSCS books stressed evolution. Today, the BSCS is credited with "putting evolution back into the biology curriculum." The BSCS books were an agent in the U.S. Supreme Court's ruling that laws banning the teaching of human evolution are unconstitutional (i.e., *Epperson v. Arkansas*), as well as in cases involving issues such as instruction about human reproduction and the use of live animals in biology classrooms. Some states, such as Texas (in 1970) and Kentucky (in 1965), banned the BSCS books. Evangelist Reuel Lemmons of Austin condemned the textbooks as the "most vicious attack we have ever seen on the Christian religion."

Today, the BSCS is a nonprofit entity headquartered in Colorado Springs, Colorado, and continues to publish high-quality biology textbooks; domestic sales of BSCS *Green* have exceeded 2.6 million copies, and sales of *Blue* have exceeded 1.6 million copies. BSCS materials have been printed in more than 25 languages for use in more than 60 countries. Since its inception, more than 20 million students have used BSCS materials.

BIBLE CRUSADERS OF AMERICA (Est. 1925)

Bible Crusaders of America continued William Jennings Bryan's crusade against the teaching of human evolution in public schools. The Crusaders were funded by real estate tycoon George Washburn, a friend of Bryan. Washburn enlisted virtually all of the leading antievolutionists for his cause, including John Straton and William Riley. The Crusaders' campaign director was T. T. Martin, who helped convince the Mississippi legislature to pass a law banning the teaching of human evolution. The Crusaders promoted the work of creationists such as Arthur Brown (1875–1947), a surgeon who claimed that evolution—"the greatest hoax ever foisted on a credulous world"—was a tool used by Satan to attack the Bible.

GREGOR MENDEL (1822–1884)

I knew that the results I obtained were not easily compatible with our contemporary scientific knowledge, and that under the circumstances publication of one such isolated experiment was doubly dangerous; dangerous for the experimenter and for the cause he represented.

Johann Mendel was born July 22, 1822, in the village of Hyncice in Moravia (now part of the Czech Republic). Johann's father was a farmer who hoped his only son would help work the farm as an adult. However, Johann's intellectual abilities and interests in science encouraged the parish priest to suggest that Johann should attend a larger school in a neighboring town. Even though Mendel's parents were of modest means, they agreed. Academic success allowed Johann to move to the Gymnasium in Troppau, from which he graduated in 1840. He was by this time supporting himself financially.

Mendel then entered the University of Olmütz, where he focused on mathematics and physics. After financial problems forced Mendel to leave the university, he entered the Augustinian Order at St. Thomas Monastery in Brünn, the capital of Moravia. The depth of Mendel's faith has never been questioned, but his decision to become a priest was likely influenced by his interests in science. The city of Brünn was then a center of cultural and scientific activity, and the abbot of the monastery supported scientific inquiry, especially as it related to agriculture. These features undoubtedly made the monastery at Brünn attractive to Mendel; he joined the monastery (and was given the name Gregor) in 1843 and was ordained in 1847.

To his disappointment, Mendel discovered that he was not well-suited to his chosen career and found visiting the sick and dying—a standard duty for a priest—particularly distressing. Mendel was moved to a teaching position, which required him to pass a certification examination, which he failed. He spent 1851 to 1853 at the University of Vienna in a teacher's training program, taking additional coursework in mathematics and science, including botany. Mendel returned to the monastery and again failed the exam. He was then allowed to serve as a substitute teacher in the local secondary school until he was elected abbot 14 years later.

In 1854, Mendel was provided space in the monastery's garden to conduct experiments on hybridization in garden peas, *Pisum sativum*. The area surrounding Brünn was agricultural, and because the monastery was a regional center of scientific study, this project to understand the process of hybridization could benefit local farmers. Mendel studied the inheritance of seven different traits in peas, with each trait studied alone or in combination with other traits. Mendel, a meticulous experimentalist and record-keeper, generated large data sets (i.e., he frequently examined hundreds of individuals of each generation) that allowed him to identify the frequencies of the alternate forms of the traits he studied. He was therefore able to discern patterns that demonstrated much about the basis of heredity; if he had relied only on qualitative data (e.g., presence/absence), he would have been unable to reach the conclusions he did. Furthermore, Mendel was comfortable applying mathematical models to explain his data, an unusual skill for a natural historian at that time, which reflected his training in mathematics and physics.

In 1865, Mendel reported his findings at two presentations to the Natural Science Society of Brünn, which met at 22 Jánská Street; that building, which is marked with a commemorative plaque, is now a technical college. The following year, a paper ("Experiments in Plant Hybridization") describing Mendel's results was published in the Society's *Proceedings*. (Mendel published only four papers during his brief research career. Besides the 1866 paper, he published one in 1854 on damage to plants by pea beetles, one in 1870 on breeding experiments in hawkweed, and one in 1871 on a tornado that caused extensive damage in Brünn.) His results revealed several important aspects of the nature of heredity. First, he demonstrated that the "potentially formative elements" (later called *genes* by Wilhelm Johannsen) act like particles by maintaining their integrity across generations, rather than being blended together as predicted by the "blending" model of heredity. Second, Mendel showed that genes exist in different forms (with different forms producing different versions of a trait, e.g., green seeds versus yellow seeds); these alternate forms of a gene are now called *alleles* (i.e., there are two different versions of the gene that determines seed color, one that codes for green, and one that codes for yellow). Third, individuals have two copies of each gene, and these copies segregate into separate cells during the production of sex cells (gametes). This *principle of segregation* predicts that for individuals having two different copies of the same allele (the heterozygous condition), half of the sex cells should contain one allele, and half should contain the other allele. Finally, when examining two traits (e.g., seed color and seed texture) simultaneously, where each trait is determined by a separate gene, Mendel determined that the probability of finding a specific version of one trait (e.g., yellow seeds) is independent of inheriting a specific version of the other trait (e.g., wrinkled seeds). This *principle of independent assortment* applies to genes located on different pairs of chromosomes (i.e., genes that are not linked).

Mendel's results did not immediately create interest. His 1865 presentation was attended by only 40 people, and apparently little discussion followed his talk. His 1866 paper likewise did not attract attention, as most readers incorrectly interpreted his conclusions as merely confirming that hybridization eventually leads to reversion to the ancestral form. However, after publication of *Origin of Species*, the issue of heredity became of greater interest, and in 1900, Mendel's results were "rediscovered" by Hugo de Vries and others. For a while, "Mendelism" was synonymous with de Vries' mutational model of evolution. As Theodosius Dobzhansky has noted, "some log jams had to be cleared before Darwinism and Mendelism could join forces," but by the early 1900s, these obstacles were being removed, principally by Thomas Hunt Morgan's work demonstrating that genes were located on chromosomes. This set the stage for the revolution of the 1920s and 1930s known as the modern synthesis.

Reexamination of Mendel's results has speculated that Mendel's frequency data fit the model of inheritance he identified exceedingly well. Ronald Fisher concluded that Mendel's results had been faked, although he blamed an assistant of Mendel's who may have independently "helped" the monk achieve the anticipated results. Others, however, have suggested that if the sample sizes Mendel used varied from the average values he reported, this could account for the close fit between the expected and observed frequencies. Regardless, there is no evidence that Mendel purposely conspired to collect his data in a biased manner, or that he adjusted his data after collection.

Mendel hoped to replicate his results from peas in another species, and he began studying hybridization in hawkweed (*Hieracium*). Unfortunately, these attempts were frustrated because, unknown to Mendel, hawkweeds reproduce asexually as well as sexually, which meant that the frequencies of traits among offspring were due not only to sexual reproduction between the individuals he crossed. This frustration, combined with his increased administrative duties when he was elected abbot in 1868, ended Mendel's research into hybridization and heredity. However, he did continue studies of meteorology and beekeeping.

The disappointment of his research program on hybridization, combined with ongoing tension between himself and the Austrian government over issues of taxation of the monastery, placed considerable strain on Mendel during the latter part of his life. Mendel, who suffered from kidney problems, died on January 6, 1884, and is buried in the monastery's cemetery in Brünn, near a plaque noting that Mendel "discovered the laws of heredity in plants and animals. His knowledge provides a lasting scientific basis for recent progress in genetics."

Today, Mendel is commemorated with a museum (the Mendelinium) at his Augustinian monastery that includes Mendel's notebooks, frescos of his plants, and the names of scientists punished during the Lysenko era. Outside the monastery are Mendel's garden (which grows plants in rows labeled P_1, F_1, and F_2), the foundations of his greenhouse, and a large statue—commemorated in 1910—of Mendel and his pea plants (in the 1950s, the statue was hidden from Communists, who wanted to destroy it). Above the entrance to Mendel's museum is the inscription, "My time will come."

KEN MILLER (b. 1948)

Any religious person who is astounded by the cruelty that we see in the world has to find some way to account for the presence of a knowing and loving God alongside that cruelty. I actually think that evolutionary biology helps a Christian to account for that in a remarkable way. Evolutionary biology shows that all life is interrelated and that life, unfortunately, only comes at the expense of death.

Kenneth Raymond Miller was born in Rahway, New Jersey, on July 14, 1948. He was raised a Catholic, a faith he has adhered to for most of his life (as a child, he even considered becoming a priest). During the summer between high school and the start of college, while working as a lifeguard, Miller read several books, including *Origin of Species*. Although Darwin's ideas had little immediate effect on Miller, it was clear that "people were afraid of this book," including his father, who cautioned his son that the book was dangerous and to "be careful not to lose your values." The basis for these warnings was not apparent to Miller, who found nothing controversial in a book that discussed seemingly well-established facts.

In the fall of 1966, Miller entered Brown University and began to question his religious faith. As an aspiring poet (he claims his efforts "occasionally rose to the heights of mediocrity"), he read widely, and came across work by poet and Trappist monk Thomas Merton. Merton's autobiography *The Seven Storey Mountain* (1948), which described his conversion to Roman Catholicism, struck "a resonant chord" with Miller and convinced him that it was possible to be an intelligent person of faith. Miller's faith was reaffirmed and has remained strong ever since.

Miller graduated from Brown with a biology degree in 1970. He then went to the University of Colorado to study cell biology, earning his Ph.D. in 1974. That same year, he became a lecturer at Harvard University and head of the electron microscopy lab, becoming an assistant professor in 1976. In 1980, he returned to Brown University as an assistant professor, rising to the rank of professor in 1986. Miller's research has focused on photosynthesis, particularly the functioning of photosynthetic membranes. He has edited *The Journal of Cell Biology* and *The Journal of Cell Science*.

During Miller's second semester at Brown, a group of students asked him to debate Young-Earth creationist Henry Morris of the Institute for

Creation Research. Miller initially declined, citing the fact that he was not an evolutionary biologist and had never heard of Henry Morris. The students (probably aware of how to goad their professor into agreeing) asked if that meant that Morris was right about evolution. Miller shot back an emphatic "no," and agreed to consider debating Morris after reviewing recordings of Morris' previous debates. In so doing, Miller quickly realized that Morris was a skilled debater who could best eminent scientists through use of "tiny little arguments" that purported to show evolution as a flawed idea. Miller realized that Morris could easily sway a naive audience and he decided that the debate was important enough to require his participation.

Miller spent a month reading, talking with colleagues, and reviewing Morris' earlier debates. Armed with two carousels of slides, Miller confronted Morris in front of nearly 3,000 spectators in April 1981. Unlike in previous debates, Morris faced an opponent who had not taken him lightly, and one that went on the offensive instead of merely dismissing creationists' claims. Afterward, Morris acknowledged Miller's abilities and claimed that he was "deviously sophisticated in argumentation" and "the best evolutionist debater to surface to date." Both creationists and evolutionary biologists noticed Miller's success, and additional debates between Morris and Miller soon followed. Miller also participated in debates with Morris' ICR colleague Duane Gish, as well as intelligent-design advocate Michael Behe.

Miller testified in *Selman v. Cobb County School District*, a case that tested whether the Cobb County, Georgia, school board could insert "warning" stickers about evolution in high school biology textbooks. Miller's textbook, coauthored with Joseph Levine, would receive the stickers, and Miller was called to discuss the content and purpose of the book. Miller was also an expert witness for the plaintiffs in the *Kitzmiller v. Dover Area School District* trial over the constitutionality of a statement to be read to ninth-grade biology students that offered intelligent design as an alternative to evolution. (As in Cobb County, the Dover school district was also using Miller's textbook.) Because Miller was an articulate defender of evolution and a devout Christian, he was free of the charge of atheism that antievolutionists frequently level against scientists, and was therefore an ideal authority to speak on behalf of the plaintiffs.

Miller tackled the relationship between science and religion in his 1999 book, *Finding Darwin's God: A Scientist's Search for Common Ground Between God and Evolution*. Miller, an excellent writer, used the first two-thirds of the book to discuss and refute the various forms of antievolutionist thought, ranging from Young-Earth creationism to intelligent design. This section of the book has been consistently reviewed positively, especially by biologists. The remainder of the book melds Miller's perspective as a scientist with his personal religious beliefs; this section has received significantly more criticism, especially from the scientific community. Miller concluded that God created a universe based on contingency rather than one that is deterministic and preordained, and natural processes are the products of that divine decision. Miller claimed that "given evolution's ability to adapt, to innovate, to test, and to experiment, sooner or later it would have given the Creator exactly what He was looking for." These conclusions elicited criticism that Miller was invoking teleology and design, aspects of nature he seemed to have refuted in earlier chapters.

Finding Darwin's God catapulted Miller into the media spotlight. He has discussed God and evolution on a variety of television programs, ranging from William F. Buckley's *Firing Line* to Comedy Central's conservative talk show parody *The Colbert Report*. He also umpires NCAA (National Collegiate Athletic Association) fast-pitch softball games.

STEPHEN JAY GOULD (1941–2002)

Humans are not the end result of predictable evolutionary progress, but rather a fortuitous cosmic afterthought, a tiny little twig on the enormously arborescent bush of life, which if replanted from seed, would almost surely not grow this twig again.

Stephen Jay Gould was born September 10, 1941, in Queens, New York. His career path was determined at the age of five when, awestruck at the sight of a *T. rex* skeleton at the American Museum of Natural History (AMNH), he decided to become a paleontologist. As a boy, Gould read everything he could about dinosaurs, and his penchant for collecting fossils earned him the name "Fossilface" from his peers. His parents encouraged his scientific pursuits, especially his father, whom Gould described as "an intellectual without official credentials." (Gould's father was also a Marxist, which may explain the younger Gould's leftward political leanings, sometimes evident in his writing.) Upon entering Jamaica High School in Queens, he eagerly anticipated learning more about evolution (Charles Darwin and Joe DiMaggio—Gould was a lifelong New York Yankees fan—were his childhood idols). But Gould was quickly dismayed at how little the subject was discussed, a situation he later attributed to the lasting effects of the Scopes trial.

Gould earned a B.A. in geology from Antioch College in Ohio in 1963, after which he entered graduate school at Columbia University to study with Norman Newell, professor of paleontology and curator at the AMNH. He married Deborah Lee in 1965 and the following year accepted a faculty position in the Geology Department at his undergraduate alma mater. He earned a Ph.D. in 1967 from Columbia University, and the same year he joined the faculty of Harvard University, eventually becoming curator of invertebrate paleontology and Alexander Agassiz Professor of Zoology. He would remain at Harvard until his death in 2002. Gould had two sons, Jesse and Ethan. In 1995, Gould—by then divorced—married Rhonda Shearer, a sculptor and art historian.

Gould's professional career encompassed two overlapping yet distinct roles: academic scientist and popular science writer. He began his long-term study of land snails of the genus *Cerion* in the Caribbean while still an undergraduate and continued this work for years after going to Harvard, producing hundreds of scientific papers. Early on, Gould was also interested in the relationship between the development of individuals (ontogeny) and they way in which the species to which they belong evolves. His book *Ontogeny and Phylogeny* (1977) explored this subject in detail, and in general outline predicted the discipline known as "evo-devo," the interface between developmental and evolutionary biology. But it was his broader views about evolution that produced the greatest effect on the academic world, views that spilled over into the public arena.

The scientific idea most closely associated with Stephen Jay Gould is punctuated equilibrium, first introduced in print with fellow paleontologist Niles Eldredge in 1972. (Gould was a strong proponent of punctuated equilibrium but gave Eldredge primary credit as the main developer of the idea.) Gould and Eldredge were struck by how the fossil record suggests that species often remain unchanged for millions of years but that these periods of stasis can be interrupted by bursts of diversification. Building upon Ernst Mayr's idea that rapid speciation is favored in peripheral, isolated populations, Gould and Eldredge argued that the fossil record demonstrates that evolutionary change is generally nonexistent in species composed of large populations due to the homogenizing effect of interbreeding among individuals. Instead, significant evolutionary change is more likely where populations are fragmented, as on the edges of species' ranges, due to the increased strength of evolutionary forces in small, isolated populations. According to Gould and Eldredge, the lack of smooth transitional sequences in the fossil record does not, therefore, represent the unavoidably imperfect (due to the chancy process of fossilization) chronicle of slow gradual change but rather accurately reflects the actual operation of the evolutionary process.

The standard model for evolutionary change, phyletic gradualism, proposes that evolution is a slow, gradual process apparent across an entire species as that species evolves in response to changes in its environment. Because Gould and Eldredge referred to punctuated equilibrium in their 1972 paper as "an alternative to phyletic gradualism," the rapid and vocal response by the scientific community to an idea that seemingly upended conventional thinking was predictable. It is now generally accepted that both phyletic gradualism and punctuated equilibrium are represented in the fossil record. Gould and Eldredge, however, continued to maintain that punctuated equilibrium was the dominant mode for evolutionary change.

Creationists used the debate about punctuated equilibrium to question the status of evolution within science: if scientists themselves can't agree about how evolution works, then evolution cannot be the well-established concept as has been claimed. Creationists also argued that punctuated equilibrium was an example of scientists' flip-flopping (evolution is first proposed to work slowly, now it is proposed to work quickly) when confronted with contradictory evidence (i.e., lack of transitional forms in the fossil record). In particular, creationists attempted to conflate punctuated equilibrium with the discredited idea of macromutations producing large evolutionary leaps ("saltations") to question evolution.

This was, predictably, infuriating to scientists, especially to Gould, who saw his work distorted to support antievolutionism. In response, he frequently attacked creationism in his popular books and monthly "This View of Life" column in *Natural History* magazine. Collections of these columns in book form (e.g., *Ever Since Darwin, The Panda's Thumb, Hen's Teeth and Horse's Toes*) reached a wider audience, establishing Gould as a recognizable and outspoken proponent of evolutionary biology and critic of antievolutionism. In the 1970s and 1980s, Gould demonstrated how the main form of antievolutionism at that time, "creation science," was, in fact, an oxymoron. Ironically, the anti-evolution forces tried to interpret this active response as an indication of the thinness of scientists' arguments.

Gould testified for the plaintiffs in *McLean v. Arkansas Board of Education*, a lawsuit that challenged Arkansas' Act 590 mandating equal time for evolution and creationism. Even though Judge William Overton ruled Act 590 to be unconstitutional and credited Gould's testimony as being influential, Gould's tendency to provide responses that could be misinterpreted except by trained biologists ("I don't believe that mutation and natural selection are sufficient . . .") was used by creationists to portray evolution as a weak concept with little explanatory power. Inadvertently abetting the creationist agenda was a charge Gould faced during much of his career.

The triumph in Arkansas and soon after that of *Edwards v. Aguillard* (another case involving "equal time" for creation science, this time in Louisiana) encouraged Gould to mistakenly declare in 1987 that the fight against creationism was over. However, the rise of intelligent design in the 1990s prompted Gould to renew his defense of the teaching of evolution. When, in the late 1990s, the Kansas Board of Education eliminated evolution from that state's science standards, Gould testified on the need for evolution in the public schools (e.g., "to teach biology without evolution is like teaching English without grammar"). Gould saw creationism as "politics, pure and simple" and part of the wider effort of religious conservatives to push that overall agenda.

The continued conflict between some religious groups (biblical literalists) and science fascinated Gould. Although a declared agnostic and a secular humanist, he had long admired religion and had stated that he regretted not having been raised within a particular religious tradition. Gould explored these topics in his 2002 book *Rock of Ages*, in which he claimed that science and religion are independent ways of knowing about the universe ("non-overlapping magesteria" or NOMA), a position embraced by some but rejected by others. In particular, Oxford University biologist Richard Dawkins (with whom Gould had frequent disagreements about various aspects of evolutionary biology) was scathing in his indictment of Gould's acceptance of religion, chastising Gould for "a cowardly flabbiness of the intellect."

Gould's development of NOMA was consistent with his contention that morality is independent of evolution. Gould bristled at the suggestion of biological determinism of human behavior, and had long disagreed with Harvard biologist Edward O. Wilson about the genetic basis of human behavior. Wilson had founded modern sociobiology, which holds that behavior (in particular social behavior) results from the reproductive advantage of genetically determined behavioral traits (i.e., via natural selection). Gould's 1981 book *The Mismeasure of Man* dealt primarily with intelligence, a trait that had been proposed as being in large measure biologically determined, and examined the negative implications for individuals and society of such a perspective. The most recent manifestation of sociobiology, evolutionary psychology, was viewed by Gould as pseudoscience that had sinister political and social implications that amounted to a resurrection of the social Darwinism of the late nineteenth century and the eugenics programs of the early twentieth century.

Gould argued throughout his career for the crucial role of "contingency" in the evolutionary process. He often noted that humans likely exist only because of chance events: restart the process of biological evolution on Earth and our species would probably not arise. This position was frequently misstated to suggest that evolution was a random process. Instead, Gould's point was that evolution is not goal-directed (any particular species or even an increase in complexity is not inevitable) and can be influenced by chance events (e.g., an asteroid impact may have allowed mammals to diversify quickly after the demise of the dinosaurs). Gould also used this position to attack those he termed "Darwinian fundamentalists" (most notably Richard Dawkins and philosopher Daniel Dennett), who he thought wrongly viewed natural selection as the only significant evolutionary force.

Gould's ability to intrigue a reader with discussions of even arcane aspects of science attracted many students to careers in science. Gould was also an accomplished historian and philosopher of science, cited by Ronald Numbers as second only to Thomas Kuhn in influence. But many have also questioned Gould's merits as a scientist; for example, eminent evolutionary biologist John Maynard Smith noted that "the evolutionary biologists with whom I have discussed his work tend to see him as a man whose ideas are so confused as to be hardly worth bothering with." Regardless, Gould received numerous awards during his career, including a Schubert Award for excellence in paleontological research (1975), an American Book Award (1981), and a MacArthur Foundation Award (1981). He also became fellow of the American Association for the Advancement of Science (1983; serving as President in 2000) and was elected to the National Academy of Sciences (1989).

In 1982, Gould was diagnosed with abdominal mesothelioma, a rare form of cancer, usually due to asbestos exposure. At the time of his diagnosis, the median expected length of survival was eight months; however, he underwent a new experimental treatment, and lived another twenty years. But cancer eventually returned, although in a different form, and Gould died in 2002.

WILLIAM JENNINGS BRYAN (1860–1925)

The contest between evolution and Christianity is a duel to the death. If evolution wins, Christianity goes.

William Jennings Bryan was born on March 19, 1860, in Salem, Illinois, the same town in which young John Scopes would later attend high school and learn biology. Bryan's wealthy parents were devout Christians.

After graduating with honors from Illinois College in 1881 and studying law at Union College of Law in Chicago, Bryan was admitted to the bar in 1883. The following year he married Mary Baird, who was admitted to the bar in 1888 but never practiced. When John Scopes graduated from high school in Salem, Bryan delivered the commencement address.

Bryan practiced law in Illinois, and then in Nebraska, before being elected to the U.S. Congress in 1890. In Congress, Bryan advocated a variety of progressive causes, including a graduated income tax, women's suffrage, and the free coinage of silver. After losing a bid for the Senate in 1894, Bryan edited the *Omaha World-Herald* and was a popular public speaker. Many of Bryan's speeches stressed his belief that the dollar should be backed by silver rather than gold. Bryan, who greatly admired Thomas Jefferson, often urged Christians to solve the problems created by "the arrogance of wealth."

When the 36-year-old Bryan went to the Democratic National Convention in Chicago in 1896, he was not widely known. The Convention deadlocked on a presidential nominee for four ballots, at which time Bryan rose to speak. There, late in the evening of July 9, Bryan defied his party's conservative leader (Grover Cleveland) with his magnificent "Cross of Gold" speech: "You shall not press down upon the brow of labor this crown of thorns, you shall not crucify mankind upon a cross of gold." The speech electrified the convention, and Bryan was nominated for president. At the time, "Boy Bryan" was the youngest person ever nominated for the presidency. During the campaign, Bryan—who was outspent 20-to-1 by his Republican opponent William McKinley—traveled almost 20,000 miles, and was the first presidential candidate to take his message directly to voters, often from the back of railroad cars (this was a new tactic, since presidential candidates had traditionally stayed home and let others speak on their behalf). Although Bryan gave 600 speeches in 27 states during the campaign, McKinley defeated Bryan by an electoral vote of 271–176 and by a popular vote of 51–47 percent. (Many other Democratic candidates for national office lost during that election, including Clarence Darrow.) In 1900, Bryan—nicknamed "The Great Commoner" because of his faith in the goodness of common people—lost again to McKinley, and in 1908 he lost to William Taft. Throughout his life, Bryan's belief in the majority remained strong; as he often asked, "By what logic can the minority demand privileges that are denied to the majority?"

In 1898, Bryan served with a Nebraska regiment in the Spanish-American War. When Woodrow Wilson became president in 1912 (after a convention battle that blocked Bryan's fourth nomination), Wilson appointed Bryan as Secretary of State. Bryan had promised that "there will be no war while I am Secretary of State," and resigned on June 9, 1915, over "war preparedness" that led the United States into World War I.

In 1901, Bryan founded *Commoner*, a weekly newspaper that he published for twelve years. At the height of its popularity, *Commoner* had more than 140,000 subscribers. Bryan continued to advocate social reforms such as prohibition, all of which were based on Bryan's deep religious faith. Bryan did not separate politics and religion, and his policies were often described as "applied Christianity." At Bryan's Sunday school classes, which attracted thousands of worshippers, Bryan often attacked evolution, claiming that "more of those who take evolution die spiritually than die physically from smallpox."

As a young man, Bryan had investigated evolution and decided "to have nothing to do with it." However, in 1916, Bryan was alarmed by *The Belief in God and Immortality*, in which Bryn Mawr psychology professor James Leuba showed that most scientists were nonbelievers and that college eroded students' religious faith. Bryan was especially troubled by human evolution, dismissing the rest because it "does not affect the philosophy upon which one's life is built." Interestingly, all laws banning the teaching of evolution banned only the teaching of *human* evolution.

By 1920, Bryan had labeled evolution "the most paralyzing influence with which civilization has had to contend during the last century." Bryan, who had little use for theistic evolution ("it deadens the pain while the Christian's religion is being removed"), began promoting the evils of evolution, and his pamphlet *The Menace of Darwinism* was distributed nationwide. Bryan claimed that evolution was merely "a guess," that "not one syllable in the Bible" supports evolution, and that "neither Darwin nor his supporters have been able to find a fact in the universe to support their hypothesis." Bryan also claimed that science must bow before religion; as he often noted, "If the Bible and the microscope do not agree, the microscope is wrong." Bryan toured the country proclaiming his message, often appearing with other antievolution crusaders such as Billy Sunday, Frank Norris, and William Riley. In response, Reverend Harry Emerson Fosdick told reporters that "the real enemies of the Christian faith are not the evolutionary biologists, but folks like Mr. Bryan who insist on setting up artificial adhesions between Christianity and outgrown scientific opinions."

In 1924, Bryan—seeking to strengthen his scientific credentials—joined the American Association for the Advancement of Science, and the following year he delivered his famed "Is the Bible True?" speech in Nashville. Soon thereafter, John Butler introduced legislation to ban the teaching of human evolution in Tennessee's public schools. When the Butler Law was passed, Bryan telegrammed Governor Austin Peay that "The Christian parents of the State owe you a debt of gratitude for saving their children from the poisonous influence of an unproven hypothesis." Bryan's speeches drew huge crowds.

On May 12, 1925—just five days after John Scopes was "arrested" in Dayton for teaching human evolution—Riley asked Bryan to represent the World's Christian Fundamentals Association in Scopes's upcoming trial. When Bryan reached his next stop in Pittsburgh, Pennsylvania, Bryan wired Riley that he would "be pleased to act for your great religious organizations and without compensation." Sue Hicks, a local attorney in Dayton on the prosecution team, told Bryan, "we will consider it a great honor to have you with us in this prosecution." Although Bryan was not as theologically rigid as many fundamentalists (he did not have *The Fundamentals* in his home library), he looked forward to the trial, and thanked Riley for "the opportunity the Fundamentalists have given me to defend the faith."

Bryan asked several other fundamentalist leaders to come to Dayton to assist him. However, flood geologist George Price was in London, and J. Gresham Machen politely declined Bryan's request, as did Sunday. John Stratton promised to attend but never showed up. Frank Norris also promised to be there, but instead sent a stenographer, and William Riley went to Seattle to fight modernists at the Northern Baptist Convention. The absence of these fundamentalists may have explained Bryan's opposition to expert witnesses for the defense (since he had none to offer for the prosecution). Although Bryan's allies did not come to Dayton, Bryan's entry into the trial made it a world-class event.

In 1923, when editorials in the *Chicago Tribune* condemned Bryan's antievolution campaign, Bryan wrote a letter to the newspaper that prompted a response from famed defense lawyer Clarence Darrow. Darrow's letter was published on the front page of the July 4, 1923, issue of the newspaper, and included more than 50 questions to which Bryan never responded. Darrow, realizing that Dayton might be his chance to confront Bryan, volunteered his services for Scopes's defense.

Throughout the pretrial publicity, Bryan stood firm about his beliefs: "We cannot afford to have a system of education that destroys the religious faith of our children." Bryan deplored Social Darwinism (be it by German militarists or American tycoons such as Andrew Carnegie), believing that it would lead to exploitation of workers, corruption of government, and moral collapse of the country. Bryan believed that natural selection was "the law of hate," and told Hicks that the Scopes trial would "end all controversy."

Bryan spent the first four days of the trial listening and fanning himself with a large palm leaf fan, which he claimed was evidence of "the great eternal plan of adapting all nature to man's use." Bryan then attacked Darwin's theory and ridiculed biology textbook author George Hunter for grouping humans into a category with "3,499 other mammals—including elephants!" Bryan then argued that evolution threatened morality. When the defense called their expert witnesses, Bryan proclaimed that "the Bible is not going to be driven out of this court by experts." Defense attorney Dudley Malone responded with his famed "We Are Not Afraid of the Truth" speech, which Bryan later said was "the greatest speech I ever heard."

Throughout the trial, Mary Bryan sent "bulletins" to her absent daughters. To Mary, Dayton's citizens were "mountain people" who were "pathetic," "do not shave every day," and who "marry and intermarry until the stock is very much weakened." Mary described John Scopes was "a long-jawed mountain product," Darrow as having "a weary, hopeless expression," and Hays as being "as self-asserting as the New York Jews can be."

On July 20, Bryan was confronted by Darrow in the most memorable event of the Scopes trial. Darrow's relentless questioning wore down Bryan, and Bryan's admission that "days" of Genesis might have been "periods" caused many of Bryan's followers to question Bryan's convictions. The next day, Bryan's testimony was expunged from the record because, Judge John Raulston ruled, it could not help determine Scopes's guilt or innocence. Decades later, evangelist Jerry Falwell claimed that Bryan "lost the respect of fundamentalists when he subscribed to the idea of periods of time for creation rather than twenty-four-hour days."

After the trial, Bryan remained in Dayton, during which time he pledged $50,000 to create a college in Dayton that would promote fundamentalists' ideals, and scouted possible sites for the college. Bryan then fulfilled a promise to fellow prosecutor A. T. Stewart by speaking at the county fair in Winchester, Tennessee, the home of Stewart and Judge Raulston. After having lunch with Stewart and Raulston in

Winchester, Bryan told reporters, "If I should die tomorrow, I believe that on the basis of the accomplishments of the last few weeks I could truthfully say, well done." He then returned to Chattanooga, where he was told by a physician to rest (Bryan had diabetes and a heart condition). Bryan had made arrangements with George Milton of the *Chattanooga News* to publish his much-anticipated 15,000-word closing speech (when the defense waived its closing argument, the prosecution was also barred from offering a closing argument). Bryan's speech stressed majoritarianism and his belief that evolution contradicted the Bible, destroyed faith in God, and "is not truth." Bryan replaced some of the text with the last words he would ever write, "With hearts full of gratitude to God." Bryan also wrote a letter to Frank Norris, thanking him for getting Bryan involved with the case and noting that, "Well, we won the case. It woke up the country. . . . Sorry you were not there."

On Sunday morning, Bryan drove to Dayton. He appeared in good health and spirits, noting that he would "stay [involved in Scopes's appeal] and see it through." Bryan then made his last public appearance at morning worship services at First Southern Methodist Church. At lunch, he told his wife that a doctor had examined him the previous day and found him in excellent condition, noting repeatedly that "I never felt better in my life" and that there was "nothing to worry about." After making a few telephone calls and arranging a vacation in the Smoky Mountains for the following week, the 65-year-old Bryan went upstairs at 3:00 PM for a nap. He never woke up. Mary Bryan sent William "Jimmy" McCartney (the family chauffeur) to awaken her husband at 4:30 PM so Bryan could prepare for his sermon that evening at the First Southern Methodist Church. McCartney was the first to see Bryan's corpse.

Bryan died in his sleep in the home of Richard Rogers on South Market Street, just a few blocks from the Rhea County Courthouse. Walter F. Thomison was the attending physician who signed Bryan's death certificate. Prosecutors Sue Hicks, Herbert Hicks, Wallace Haggard, and Gordon McKenzie watched Bryan's body through the night, after which an honor guard of servicemen ringed the house. The Ku Klux Klan burned crosses in Bryan's memory, and Mary Bryan received thousands of messages of condolences, including those from President Calvin Coolidge, U.S. Supreme Court Chief Justice William Taft, and several presidents of foreign countries. Coolidge, who dismissed the Scopes trial as a regional event, praised Bryan's many years of public service and ordered flags at all national buildings to be flown at half-mast. Famed evangelist Billy Sunday declared Bryan to be "God's Napoleon . . . one of the world's immortals . . . who fell with his face to the enemy." Bryan's much-anticipated *Last Message*, which was originally titled *Fighting to Death for the Bible* and touted by Bryan as "the mountain peak of my life's efforts," was distributed on July 28 as his body lay in state in Dayton. The final words of Bryan's last speech came from one of his favorite hymns: "Faith of our fathers— holy faith, we will be true to thee till death."

Many people wrote songs immortalizing Bryan, and others made Bryan a martyr; Governor Austin Peay, "The Maker of Modern Tennessee" who had signed John Butler's legislation into law, pro- claimed that Bryan had died "a martyr to the faith of our fathers" and announced a state holiday to commemorate Bryan's funeral. Fundamentalists compared Bryan to Jesus Christ, and Scopes's defend- ers to Pontius Pilate and other biblical villains. William Riley, who had been instrumental in getting Bryan into the Scopes trial, described Bryan as "the great outstanding man of our movement" and the Scopes trial as "Bryan's best and last battle." Others, however, contin- ued to attack Bryan. For example, socialist Eugene Debs, one of Darrow's former clients, announced that "the cause of human progress sustains no loss in the death of Mr. Bryan." Darrow expressed sorrow about Bryan's death and commended Bryan's ability, courage, and strong convictions, and Ben McKenzie hailed Bryan as "the noblest hero of these times." Later, however, Darrow expressed pity that Bryan had been "obliged to show his gross ignorance" and that Bryan's vanity and failures had driven him to "a state of hallucination that would impel [Bryan] to commit any cruelty that he believed would help his cause." Reporter H. L. Mencken began Bryan's obituary in the *Baltimore Sun* by asking, "Has it been duly marked by historians that William Jennings Bryan's last secular act on this globe of sin was to catch flies?"

Bryan's memorial service in Dayton was held on the lawn of the Rogers' home where he died, and was officiated by Reverend C. R. Jones, the pastor of Dayton's First Southern Methodist Church. Dayton's mayor proclaimed a day or mourning, and Dayton's flags were flown at half-mast. Crowds of mourners lined railroad tracks to see the special Pullman car that carried Bryan's body to Virginia for burial. The train's conductor had taken Bryan along a similar route during Bryan's presidential campaign in 1896.

Bryan's funeral in Washington, D.C., on July 31, was held at New York Avenue Presbyterian Church, "The Church of the Presidents," and was broadcast nationwide via radio. Bryan was buried atop a tree-covered hill in the south end of Arlington National Cemetery, where he rests with his wife Mary (1861–1930) beneath the tiny inscription, "He Kept the Faith."

Knowing that "my family history does not promise a long life," Bryan had prepared his will on July 5, 1925, just before coming to Dayton. The will, which opened with "In the name of God, farewell," was settled in early August 1925. Bryan's estate was valued at $860,500. Bryan's wife received their home in Florida (Marymont in Coconut Grove) and one-third of the remaining estate. Each of Bryan's children received $100,000. Bryan noted that he had "hoped to aid in the establishment of an academy that would embody my idea [of a Christian school]" at which "boy students [would] wear a uniform of blue and gray, to symbolize the union of north and south." Bryan left all publication rights of his writings to his widow and children, who were authorized to publish Bryan's official biography. Bryan insisted that the biography describe his wife's life, "setting forth particularly the aid which Mrs. Bryan has rendered to her husband in all his work during his life." Bryan's daughter Ruth Bryan Owen (1885–1954) later became the first woman from Florida to serve in the U.S. House of Representatives.

Soon after Bryan's death, F. E. Robinson, Judge John Raulston, A. T. Stewart, Walter White, Ben McKenzie, and others formed a memorial association to help create the college that Bryan proposed in the final days of his life. Although almost half of the $1,000,000 gathered for William Jennings Bryan University was erased by the Great Depression, ground was broken on November 5, 1926, by Governor Peay for the conservative Christian school (whose name was changed to William Jennings Bryan College in 1958, and then shortened to Bryan College in 1993). The ceremony was attended by more than 10,000 onlookers, and Bryan College opened in 1930 in the old high school building where John Scopes allegedly taught evolution. The first class included 31 students, and by 1939–1940, the college enrolled ninety students. Today, Bryan College is a coeducational liberal arts college located atop a hill overlooking Dayton. The college enrolls about 600 students and is based on "Christ Above All" and the "unequivocal acceptance of the inerrancy and authority of the Scriptures." Scopes trial instigator George Rappleyea wanted to create a liberal college to offset Bryan College, but that college was never built.

After Bryan's death, several people (including New York fundamentalist John Straton) claimed Bryan's place in the antievolution crusade, but none ever matched Bryan's stature or credibility. Although religious conservatives such as George Price, Henry Morris, and Jerry Falwell have criticized Bryan's performance at Dayton, most of

Bryan's life served the public. Indeed, few statesmen have been more vindicated by history than Bryan, for many of Bryan's political causes were subsequently enacted into law, including the Sixteenth (graduated income tax), Seventeenth (direct election of senators), Eighteenth (prohibition of liquor), and Nineteenth (woman suffrage) Amendments to the Constitution, various labor laws (e.g., eight-hour workday, minimum wage), the Federal Reserve Act, and tariff reform.

In 1986, Bryan was commemorated on a $2 U.S. postage stamp, and today he is memorialized with a variety of museums (e.g., the William Jennings Bryan Birthplace Museum in Salem, Illinois; Fairview, the Bryan Museum in Lincoln, Nebraska), statues (e.g., including one on the Rhea County Courthouse lawn in Dayton, and another that was designed by famed sculptor Gutzon Borglum, who sculpted the presidents on Mount Rushmore), libraries, schools, and parks throughout the United States. Borglum's 12-foot-high statue of Bryan was dedicated in 1934 by President Theodore Roosevelt in Washington, D.C., and today stands in Salem, Illinois.

—*Randy Moore and Mark Decker*

APPENDIX B

Timeline of Major Court Cases

1925   *State of Tennessee v. John Thomas Scopes*

In the original "Trial of the Century," coach and substitute science teacher John Scopes was convicted of the misdemeanor of teaching human evolution in a public school in Tennessee. Scopes's trial, which William Jennings Bryan described as "a duel to the death" between evolution and Christianity, remains the most famous event in the history of the evolution-creationism controversy. In 1960, the Scopes "monkey" trial also provided a framework for the movie and play *Inherit the Wind*.

1927   *John Thomas Scopes v. State of Tennessee*

The Tennessee Supreme Court upheld the constitutionality of a Tennessee law forbidding the teaching of human evolution but urged that Scopes' conviction be set aside. This decision ended the legal issues associated with the Scopes trial, and the ban on teaching human evolution in Tennessee, Mississippi, and Arkansas remained unchallenged for more than 40 years:

> The last serious criticism made of the Act is that it contravenes the provision of section 3 of article 1 of the

Constitution, "that no preference shall ever be given, by law, to any religious establishment or mode of worship."

The language quoted is a part of our Bill of Rights, was contained in our first Constitution of the state adopted in 1796, and has been brought down into the present Constitution. At the time of the adoption of our first Constitution, this government had recently been established and the recollection of previous conditions was fresh. England and Scotland maintained State churches as did some of the Colonies, and it was intended by this clause of the Constitution to prevent any such undertaking in Tennessee.

We are not able to see how the prohibition of teaching the theory that man has descended from a lower order of animals gives preference to any religious establishment or mode of worship. So far as we know, there is no religious establishment or organized body that has in its creed or confession of faith any article denying or affirming such a theory. So far as we know, the denial or affirmation of such a theory does not enter into any recognized mode of worship. Since this cause has been pending in this court, we have been favored, in addition to briefs of counsel and various *amici curiae*, with a multitude of resolutions, addresses, and communications from scientific bodies, religious factions, and individuals giving us the benefit of their views upon the theory of evolution. Examination of these contributions indicates that Protestants, Catholics, and Jews are divided among themselves in their beliefs, and that there is no unanimity among the members of any religious establishment as to this subject. Belief or unbelief in the theory of evolution is no more a characteristic of any religious establishment or mode of worship than is belief or unbelief in the wisdom of the prohibition laws. It would appear that members of the same churches quite generally disagree as to these things.

1968   *Epperson v. Arkansas*

The U.S. Supreme Court struck down an Arkansas law making it illegal to teach human evolution. As result of this decision, all laws banning the teaching of human evolution in public schools were overturned by 1970:

In the present case, there can be no doubt that Arkansas has sought to prevent its teachers from discussing the theory of evolution because it is contrary to the belief of some that the Book of Genesis must be the exclusive source of doctrine as to the origin of man. No suggestion has been made that Arkansas' law may be justified by considerations of state policy other than the religious views of some of its citizens. It is clear that fundamentalist sectarian conviction was and is the law's reason for existence. Its antecedent, Tennessee's "monkey law," candidly stated its purpose: to make it unlawful "to teach any theory that denies the story of the Divine Creation of man as taught in the Bible, and to teach instead that man has descended from a lower order of animals." Perhaps the sensational publicity attendant upon the Scopes trial induced Arkansas to adopt less explicit language. It eliminated Tennessee's reference to "the story of the Divine Creation of man" as taught in the Bible, but there is no doubt that the motivation for the law was the same: to suppress the teaching of a theory which, it was thought, "denied" the divine creation of man.

Arkansas' law cannot be defended as an act of religious neutrality. Arkansas did not seek to excise from the curricula of its schools and universities all discussion of the origin of man. The law's effort was confined to an attempt to blot out a particular theory because of its supposed conflict with the Biblical account, literally read. Plainly, the law is contrary to the mandate of the First, and in violation of the Fourteenth, Amendment to the Constitution.

1973   *Wright v. Houston Independent School District*

The Fifth Circuit Court of Appeals ruled that (1) the teaching of evolution does not establish religion, (2) there is no legitimate state interest in protecting particular religions from scientific information "distasteful to them," and (3) the free exercise of religion is not accompanied by a right to be shielded from scientific findings incompatible with one's beliefs:

Contrary to the sincere, able, and vigorous arguments of plaintiffs, the Federal courts cannot by judicial decree do that which the Supreme Court has declared the state legislatures powerless to do, i.e., prevent teaching the theory of evolution in public school for religious reasons. . . . To require the teaching of every theory of human origin, as alternatively suggested by plaintiffs, would be an unwarranted intrusion into the authority of public school systems to control the academic curriculum.

1975   *Daniel v. Waters*

The Sixth Circuit Court of Appeals overturned the Tennessee law requiring equal emphasis on evolution and the Genesis version of creation:

First, the statute requires that any textbook which expresses an opinion about the origin of man "shall be prohibited from being used" unless the book specifically states that the opinion is "a theory" and "is not represented to be scientific fact." The statute also requires that the Biblical account of creation (and other theories of creation) be printed at the same time, with commensurate attention and equal emphasis. As to all such theories, except only the Genesis theory, the textbook must print the disclaimer quoted above. But the proviso in Section 2 would allow the printing of the Biblical account of creation as set forth in Genesis without any such disclaimer. The result of this legislation is a clearly defined preferential position for the Biblical version of creation as opposed to any account of the development of man based on scientific research and reasoning. For a state to seek to enforce such a preference by law is to seek to accomplish the very establishment of religion which the First Amendment to the Constitution of the United States squarely forbids.

1977   *Hendren v. Campbell*

The County Court in Marion, Indiana, ruled that it is unconstitutional for a public school to adopt creationism-based biology books because these books advance a specific religious point of view:

Clearly, the purpose of *A Search for Order in Complexity* is the promotion and inclusion of fundamentalist Christian doctrine in the public schools. The publishers themselves admit that this text is designed to find its way into the public schools to *stress* Biblical Creationism. The court takes no position as to the validity of either evolution or Biblical Creationism. That is not the issue. The question is whether a text obviously designed to present *only* the view of Biblical Creationism in a favorable light is constitutionally acceptable in the public schools of Indiana. Two hundred years of constitutional government demand that the answer be *no*. The asserted object of the text to present a balanced or neutral argument is a sham that breaches that "wall of separation" between church and state voiced by Thomas Jefferson. Any doubts of the text's fairness is dispelled by the demand for "correct" Christian answers demanded by the *Teacher's Guide*. The prospect of biology teachers and students alike, forced to answer and respond to continued demand for "correct" fundamentalist Christian doctrines, has no place in the public schools. The attempt to present Biblical Creationism as the only accepted scientific theory, while novel, does not rehabilitate the constitutional violation.

1980   *Crowley v. Smithsonian Institution*

The D.C. Circuit Court of Appeals ruled that the federal government can fund public exhibits that promote evolution. The government is not required to provide money to promote creationism:

Application of the Supreme Court's caution to this case necessarily requires a balance between appellants' freedom to practice and propagate their religious beliefs in creation without suffering government competition or interferences and appellees' right to disseminate, and the public's right to receive, knowledge from government, through schools and other institutions such as the Smithsonian. This balance was long ago struck in favor of diffusion of knowledge based on responsible scientific foundations, and against special constitutional protection of religious believers from the competition generated by such knowledge diffusion.

1981   *Segraves v. State of California*

The Sacramento Superior Court ruled that the California State Board of Education's Science Framework, as qualified by its anti-dogmatism policy, sufficiently accommodates the views of Segraves, contrary to his claim that discussions of evolution prohibit his and his children's free exercise of religion. The state's anti-dogmatism policy specified that science discussions focus on "how" and that speculation about origins not be presented dogmatically:

> Defendant State Board of Education has acted in good faith and has taken no action which would deny plaintiffs herein and particularly plaintiff Kasey Segraves their or his rights to free exercise of their religion guaranteed to them and him by the First and Fourteenth Amendments to the Constitution of the United States and Article 1, Section 4 of the California Constitution.
>
> .... To the contrary, defendant State Board of Education has had for a number of years and currently has a policy ... which provides that in a discussion of origins in science texts and classes (a) dogmatism be changed to conditional statements where speculation is offered as explanation for origins and (b) that science emphasizes "how" and not "ultimate cause" for origins.

1982   *McLean v. Arkansas Board of Education*

An Arkansas federal district court ruled that creation science has no scientific merit or educational value as science. Laws requiring equal time for "creation science" are unconstitutional:

> The defendants presented Dr. Larry Parker, a specialist in devising curricula for public schools. He testified that the public school's curriculum should reflect the subjects the public wants in schools. The witness said that polls indicated a significant majority of the American public thought creation science should be taught if evolution was taught. The point of this testimony was never placed in a legal context. No doubt a sizeable majority of Americans believe in the concept of a Creator or, at least, are not opposed to the concept and see nothing wrong with teaching school children the idea.
>
> The application and content of First Amendment principles are not determined by public opinion polls or by a

majority vote. Whether the proponents of Act 590 constitute the majority or the minority is quite irrelevant under a constitutional system of government. No group, no matter how large or small, may use the organs of government, of which the public schools are the most conspicuous and influential, to foist its religious beliefs on others.

1987 *Edwards v. Aguillard*

The U.S. Supreme Court overturned the Louisiana law requiring public schools that teach evolution to also teach "creation science," noting that such a law advances religious doctrine and therefore violates the First Amendment's establishment of religion clause:

> In this case, the purpose of the Creationism Act was to restructure the science curriculum to conform with a particular religious viewpoint. Out of many possible science subjects taught in the public schools, the legislature chose to affect the teaching of the one scientific theory that historically has been opposed by certain religious sects. As in *Epperson*, the legislature passed the Act to give preference to those religious groups which have as one of their tenets the creation of humankind by a divine creator. The "overriding fact" that confronted the Court in *Epperson* was "that Arkansas' law selects from the body of knowledge a particular segment which it proscribes for the sole reason that it is deemed to conflict with . . . a particular interpretation of the Book of Genesis by a particular religious group." Similarly, the Creationism Act is designed either to promote the theory of creation science which embodies a particular religious tenet by requiring that creation science be taught whenever evolution is taught or to prohibit the teaching of a scientific theory disfavored by certain religious sects by forbidding the teaching of evolution when creation science is not also taught. The Establishment Clause, however, "forbids alike the preference of a religious doctrine or the prohibition of theory which is deemed antagonistic to a particular dogma." Because the primary purpose of the Creationism Act is to advance a particular religious

belief, the Act endorses religion in violation of the First Amendment . . .

The Louisiana Creationism Act advances a religious doctrine by requiring either the banishment of the theory of evolution from public school classrooms or the presentation of a religious viewpoint that rejects evolution in its entirety. The Act violates the Establishment Clause of the First Amendment because it seeks to employ the symbolic and financial support of government to achieve a religious purpose. The judgment of the Court of Appeals therefore is Affirmed.

1990   *Webster v. New Lenox School District #122*

The Seventh Circuit Court of Appeals ruled that a teacher does not have a First Amendment right to teach creationism in a public school. A school district can ban a teacher from teaching creationism:

> Given the school board's important pedagogical interest in establishing the curriculum and legitimate concern with possible establishment clause violations, the school board's prohibition on the teaching of creation science to junior high students was appropriate.

1994   *Peloza v. Capistrano Unified School District*

The Ninth Circuit Court of Appeals ruled that evolution is not a religion and that a school can require a biology teacher to teach evolution:

> According to Peloza's complaint, all persons must adhere to one of two religious belief systems concerning "the origins of life and of the universe": evolutionism, or creationism . . . Thus, the school district, in teaching evolutionism, is establishing a state supported "religion." We reject this claim because neither the Supreme Court, nor this circuit, has ever held that evolutionism or secular humanism are "religions" for Establishment Clause purposes.
>
> Indeed, both the dictionary definition of religion and the clear weight of the case law are to the contrary. The Supreme Court has held unequivocally that while the belief in a divine creator of the universe is a religious

belief, the scientific theory that higher forms of life evolved from lower forms is not.

Peloza would have us accept his definition of "evolution" and "evolutionism" and impose his definition on the school district as its own, a definition that cannot be found in the dictionary, in the Supreme Court cases, or anywhere in the common understanding of the words. Only if we define "evolution" and "evolutionism" as does Peloza as a concept that embraces the belief that the universe came into existence without a Creator might he make out a claim. This we need not do. To say red is green or black is white does not make it so. Nor need we for the purposes of a 12(b)(6) motion accept a made-up definition of "evolution." Nowhere does Peloza point to anything that conceivably suggests that the school district accepts anything other than the common definition of "evolution" and "evolutionism." It simply required him as a biology teacher in the public schools of California to teach "evolution." Peloza nowhere says it required more.

1999   *Freiler v. Tangipahoa Parish Board of Education*

The Fifth Circuit Court of Appeals ruled that it is unlawful to require teachers to read aloud a disclaimer stating that the biblical view of creationism is the only concept from which students are not to be dissuaded. Such disclaimers are "intended to protect and maintain a particular religious viewpoint":

> We find that the contested disclaimer does not further the first articulated objective of encouraging informed freedom of belief or critical thinking by students. Even though the final sentence of the disclaimer urges students "to exercise critical thinking and gather all information possible and closely examine each alternative toward forming an opinion," we find that the disclaimer as a whole furthers a contrary purpose, namely the protection and maintenance of a particular religious viewpoint. In the first paragraph to be read to school children, the Tangipahoa Board of Education declares that the "Scientific Theory of Evolution . . . should be presented to inform students of the scientific concept" but that such teaching is "not intended to influence or dissuade the

Biblical version of Creation or any other concept." From this, school children hear that evolution as taught in the classroom need not affect what they already know. Such a message is contrary to an intent to encourage critical thinking, which requires that students approach new concepts with an open mind and a willingness to alter and shift existing viewpoints. This conclusion is even more inescapable when the message of the first paragraph is coupled with the statement in the last that it is "the basic right and privilege of each student to . . . maintain beliefs taught by parents on [the] . . . matter of the origin of life. . . ." We, therefore, find that the disclaimer as a whole does not serve to encourage critical thinking and that the School Board's first articulated purpose is a sham. . . .

Here, the disclaimer approved by the School Board is to be read during school hours by school teachers and explicitly encourages students to consider religious alternatives to evolution, a part of the state-mandated curriculum. [There is a] danger of students and parents perceiving that the School Board endorses religion, specifically those creeds that teach the Biblical version of creation. The benefit to religion conferred by the reading of the Tangipahoa disclaimer is more than indirect, remote, or incidental. As such, we conclude that the disclaimer impermissibly advances religion, thereby violating the second prong of the *Lemon* test as well as the endorsement test.

2000   *LeVake v. Independent School District #656*

A Minnesota state court ruled that a public school teacher's right to free speech as a citizen does not permit the teacher to teach a class in a manner that circumvents the prescribed course curriculum established by the school board. Refusing to allow a teacher to teach the alleged "evidence against evolution" does not violate the free-speech rights of a teacher:

In the instant case, Plaintiff continues to practice his religion of choice and to belong to his religion of choice. Plaintiff makes no claim that Defendants presented him with some sort of quid pro quo whereby he was to alter his religious practices or to limit expression of his religious beliefs in order to retain his biology teaching position. Rather, Plaintiff's claim boils down to assertion that Defendants were aware of his religious beliefs and improperly considered his religious beliefs to reach a conclusion that he had a conflict between his beliefs and the theory of evolution that would prevent him from being an effective biology teacher. Therefore, asserts Plaintiff, Defendants penalized him for his religious beliefs because of their assumptions about those beliefs. Defendants deny this and assert that they reassigned Plaintiff solely because it became apparent to them that Plaintiff would not teach the District's established biology curriculum. . . .

Superintendent Dixon referred in his letter to Plaintiff to statements by Plaintiff in his position paper characterizing evolution as "impossible" and stating that he would teach evolution, but with an "honest look" at the difficulties and inconsistences. Superintendent Dixon asserted that Plaintiff had made it clear that he could not teach the curriculum.

2001 *Moeller v. Schrenko*

The Georgia Court of Appeals ruled that using a biology textbook that states creationism is not science does not violate the Establishment or the Free Exercise Clauses of the Constitution:

To prove a violation of the free exercise clause, Moeller must show that use of the textbook prevents her from practicing her religion. In this case, this showing has not been made, and no substantial burden has been placed on Moeller's free exercise of religion. The use of the textbook in question in no way forces Moeller to refrain from practicing her religious beliefs. And it does not impinge on her parents' religious instruction of their daughter. As such, Moeller's free exercise of her religion has not been substantially burdened, and use of the textbook does not violate her First Amendment rights.

2005 *Selman et al. v. Cobb County School District*

The U.S. District Court for the Northern District of Georgia ruled that it is unconstitutional to paste stickers into science textbooks claiming that, among other things, "evolution is a theory, not a fact." Such stickers convey "a message of endorsement of religion" and "aid the belief of Christian fundamentalists and creationists":

> Adopted by the school board, funded by the money of taxpayers, and inserted by school personnel, the Sticker conveys an impermissible message of endorsement and tells some citizens that they are political outsiders while telling others that they are political insiders. Regardless of whether teachers comply with the Cobb County School District's regulation on theories of origin and regardless of the discussions that actually take place in the Cobb County science classrooms, the Sticker has already sent a message that the School Board agrees with the beliefs of Christian fundamentalists and creationists. The School Board has effectively improperly entangled itself with religion by appearing to take a position. Therefore, the Sticker must be removed from all of the textbooks into which it has been placed.

2005   *Kitzmiller et al. v. Dover Area School District*

The U.S. District Court for the Middle District of Pennsylvania ruled that (1) "the overwhelming evidence ... established that intelligent design (ID) is a religious view, a mere re-labeling of creationism, and not a scientific theory," and, instead, is nothing more than creationism in disguise; (2) the advocates of ID wanted to "change the ground rules of science to make room for religion"; and (3) "ID is not supported by any peer-reviewed research, data, or publications." The judge also noted the "breathtaking inanity" of the school board's policy and the board's "striking ignorance" of ID and made the following point: "It is ironic that several of [the members of the school board], who so staunchly and proudly touted their religious convictions in public, would time and again lie to cover their tracks and disguise the real purpose behind the ID Policy."

2010   *Association of Christian Schools International et al. v. Roman Stearns et al.*

The Ninth Circuit Court of Appeals affirmed the lower court's rulings in support of the University of California. The plaintiffs had challenged the University of California's determination that Christian schools' coursework, particularly creationist biology textbooks, were inconsistent with college preparatory work:

> Plaintiffs challenge Defendants' decision to reject biology courses that used *Biology: God's Living Creation* (published by A Beka) or *Biology for Christian Schools* (published by Bob Jones University) as the primary text. Around early 2003, UC Professor Barbara Sawrey reviewed these two Christian biology textbooks and concluded that they were inappropriate for use as primary texts in college-preparatory science classes.... Professor Sawrey found these texts problematic because they characterized religious doctrine as scientific evidence, included scientific inaccuracies, failed to encourage critical thinking, and took an "overall un-scientific approach to the subject matter." ... Sawrey's "judgment was based not on the fact that the textbooks contained religious references and viewpoints, but on [her] conclusion that [the texts] would not adequately teach students the scientific principles, methods, and knowledge for them to successfully study those subjects at UC." ... Plaintiff's evidence also supports Defendants' conclusion that these biology texts are inappropriate for use as the primary or sole text. Plaintiff's own biology expert, Professor Michael Behe, testified that "it is personally abusive and pedagogically damaging to de facto require students to subscribed to an idea.... Requiring a student to, effectively, consent to an idea violates [her] personal integrity. Such wrenching violation [may cause] a terrible educational outcome." (Behe Decl. 59)

Yet, the two Christian biology texts at issue commit this "wrenching violation." For example, *Biology for Christian Schools* declares on the very first page that:

(1) " 'Whatever the Bible says is so; whatever man says may or may not be so,' is the only [position] a Christian can take ..."

(2) "If [scientific] conclusions contradict the Word of God, the conclusions are wrong, no matter how many scientific facts may appear to back them."

(3) "Christians must disregard [scientific hypotheses or theories] that contradict the Bible."

2011   *C. F. et al. v. Capistrano School District*

The Ninth Circuit Court of Appeals overturned the lower court decision in which history teacher James Corbett's remark that creationism is "superstitious nonsense," among other comments, was ruled unconstitutional:

> In broaching controversial issues like religion, teachers must be sensitive to students' personal beliefs and take care not to abuse their positions of authority.... But teachers must also be given leeway to challenge students to foster critical thinking skills and develop their analytical abilities. This balance is hard to achieve, and we must be careful not to curb intellectual freedom by imposing dogmatic restrictions that chill teachers from adopting the pedagogical methods they believe are most effective. ...
> At some point a teacher's comments on religion might cross the line and rise to the level of unconstitutional hostility. But without any cases illuminating the " 'dimly perceive[d] ... line[ ] of demarcation' " between permissible and impermissible discussion of religion in a college level history class, we cannot conclude that a reasonable teacher standing in Corbett's shoes would have been on notice that his actions might be unconstitutional.... We therefore affirm the district court's decision that Corbett was entitled to qualified immunity.

> —*Randy Moore, Mark Decker, and Sehoya Cotner*

APPENDIX C

Timeline of Media Moments in the Evolution-Intelligent Design Debate

1859   Charles Lyell urged publisher John Murray to publish Charles Darwin's "important new work." Murray agreed, and on November 24—22 years after Darwin had opened his secret "Transmutation of Species" notebook—John Murray Publishing of London (which had published all of Lyell's books) released Darwin's 502-page book *On the Origin of Species by Means of Natural Selection, or The Preservation of Favoured Races in the Struggle for Life*. Murray printed 1,250 copies of *The Origin of Species*, 139 of which were distributed as promotional copies. Booksellers bought all of the 1,111 remaining copies (for 15 shillings—about £35 today—apiece) on the first day they were for sale. Darwin's five subsequent revisions of *The Origin of Species* were published in several languages and took Darwin's idea throughout the world. Unlike Lamarck's *Philosophie Zoologique*, which was purely a theoretical book, Darwin's *Origin* was an overwhelming compendium of facts. It includes only one sentence about human evolution, but that sentence could be the understatement of the nineteenth century: "Light will be thrown on the origin of man and his history." Whereas Lamarck had spoken of the "march of nature," Darwin wrote of "transformation." When Darwin sent a copy of the book to Alfred Wallace, he enclosed a note:

"God knows what the public will think." In *The Origin of Species*, Darwin 1) replaced the notion of a perfectly designed and benign world with one based on an unending, amoral struggle for existence; 2) challenged prevailing Victorian ideas about progress and perfectibility with the notion that evolution produces change and adaptation, but not necessarily progress, and never perfection; 3) offered no larger purpose in nature for humanity other than the production of fertile offspring; and 4) liberated readers from the conceit of providentially supervised special creation of each species with the argument that all life—humans included—descended from a common ancestor; that is, humans are not a special product of creation, but of evolution acting according to principles that act on other species. For Darwin, natural selection replaced divine benevolence as an explanation for adaptation. Darwin's theory was based on an ancient Earth, and he wrote that anyone not grasping Earth's antiquity "may at once close this book." *Origin* shifted thought from a foundation of untestable awe of special creation to a science-based examination of the natural world based on natural mechanisms and historical patterns. Whereas scientists before Darwin had often invoked purpose to explain biology (e.g., a particular structure was present because it was pleasing to a deity), Darwin's idea replaced purpose with function and history; a structure was there because it was (or had been) an adaptation. (After Darwin, other scientists adopted a similar approach; for example, astronomers did not invoke purpose when describing the orbits of comets.) Darwin had many defenders, most notably Harvard scientist (and evangelical Christian) Asa Gray in the United States, and Thomas Huxley in England. Throughout the uproar that followed the publication of his book, Darwin stayed at Down House; he was interested in what was happening, but stayed out of the fray. Not surprisingly, *Origin* was condemned by many religious leaders, and William Whewell, Master of Trinity College at Cambridge, refused to allow it into the college library (despite the fact that Darwin quoted Whewell prominently opposite the book's title page). Darwin, who believed that "with a good book as with a fine day, one likes it to end with a glorious sunset," closed his book with this famous paragraph:

> It is interesting to contemplate an entangled bank, clothed with many plants of many kinds ... so different

from each other, and dependent on each other in so complex a manner ... all ... produced by laws acting around us. ... Thus, from the war of nature, from famine and death, the most exalted object which we are capable of conceiving, namely, the production of the higher animals, directly follows. There is grandeur in this view of life, with its several powers, having been originally breathed into a few forms or into one; and that, whilst this planet has gone cycling on according to the fixed law of gravity, from so simple a beginning endless forms most beautiful and most wonderful have been, and are being, evolved.

John Henslow damned Darwin's book with faint praise: "a stumble in the right direction." The following year, Darwin wrote to Charles Lyell that if he were starting anew, he would use the phrase *natural preservation* rather than *natural selection*.

1861   In *Physical Geography of the Globe*, astronomer John Herschel claimed that Darwin's theory "gave no indication of the Creator's foresight," and argued that "an intelligence, guided by a purpose, must be continually in action to bias the directions of the steps of change." Darwin responded, "[T]he point which you raise on intelligent Design has perplexed me beyond measure. ... One cannot look at this Universe with all living productions & man without believing that all has been intelligently designed; yet when I look to each individual organism, I can see no evidence of this. For I am not prepared to admit that God designed the feathers in the tail of the rock-pigeon to vary in a highly peculiar manner in order that man might select variations & make a Fan-tail." This was the first use of the phrase *intelligent design* in its modern sense. Darwin was later buried beside Herschel in Westminster Abbey.

1868   Alfred Wallace, writing in *Quarterly Review*, depicted humanity as the purpose of a divinely guided evolutionary process. Darwin was appalled by Wallace's claims, but politely responded that he hoped Wallace had "not murdered too completely your own & my child. ... I differ greatly from you, and I feel very sorry for it." Wallace responded by telling Darwin that his experiences at séances had convinced him of the reality of spirit forces, and that he now believed that God or spirits must have had an important role in human evolution.

1915   British evangelist Elizabeth Reid "Lady" Hope (1842–1922) claimed in the Baptist newspaper *Watchman Examiner* that when she visited Charles Darwin just before he died, he denounced evolution and became a Christian. Although Darwin's children denied Lady Hope's story (she was never present during any of Darwin's illnesses and he never repudiated evolution), the story became a favorite of antievolutionists.

1919   In a speech titled "Brother or Brute?," William Jennings Bryan told the World Brotherhood Congress that Nietzsche had carried Darwinism to its ultimate conclusion and that Darwinism was "the most paralyzing influence with which civilization has had to contend." Bryan argued that if Darwinism were true, humans could not overcome their animal nature, and therefore all attempts at reform will be pointless. In the same year, Bryan spoke at the high school graduation ceremony in Salem, Illinois; among the graduates who met Bryan at the ceremony was John Scopes. In 1925, Bryan helped prosecute Scopes for allegedly violating the Tennessee law banning the teaching of human evolution.

1920   William Jennings Bryan delivered for the first time his famous "Menace of Evolution" speech, which decried the dangers of evolution, which he claimed is "not science at all" but "guesses strung together." Bryan argued that his goal was to protect people "from the demoralization involved in accepting a brute ancestry." The subsequent printing of Bryan's lecture in pamphlets and newspapers reached millions of readers and produced one of the antievolution movement's most famous claims—namely, that "it is better to trust in the Rock of Ages than to know the age of the rocks."

1924   Unitarian preacher and modernist Charles Francis Potter faced militant antievolutionist and fundamentalist John Roach Straton in four highly publicized debates in New York's Carnegie Hall. Descriptions of the debates in newspapers fueled the escalating evolution-creationism controversy.

1924   Worried that his children would be corrupted by the teaching of evolution in the public schools, Tennessee state legislator John Butler (1875–1952) drafted what would be known as the Butler Law (House Bill No. 185). This law made the teaching of human evolution (i.e., "Any theory that denies the story of the Divine Creation of man as taught in the Bible, and to teach instead that man has descended from a lower order of animals") in any of Tennessee's public schools a misdemeanor punishable by a fine of $100–500. Although editors of *The Chattanooga Times* urged that Butler's legislation be ignored, his proposed ban on teaching human evolution was otherwise unopposed. Butler's legislation, which became law the following year, was the basis for the Scopes trial, the most famous event in the history of the evolution-creationism controversy.

1925   John Scopes's contract with Rhea County High School expired on May 1, but four days later he volunteered to be arrested for the teaching of human evolution and was charged the following day. George Rappleyea wired the ACLU a plan for a "four-round fight" that would culminate with a hearing at the U.S. Supreme Court. Two days later, *The Washington Post* announced the story on its front page: "J. T. Scopes, of the science department of the Rhea County High School, was arrested by a deputy sheriff, charged with violating the Tennessee law prohibiting the teaching of evolution in the state public schools." Fundamentalist leader William Bell Riley asked William Jennings Bryan to represent WCFA at Scopes's trial. On May 12, Bryan—who hoped that the upcoming trial would return him to the front pages of the nation's newspapers—responded that he would "be pleased to act for your great religious organizations and without compensation." Bryan, who thanked Riley for "the opportunity the Fundamentalists have given me to defend the faith," believed the Scopes trial would "end all controversy." In Dayton, Bryan and his entourage eventually took over four rooms of the home of local druggist F. R. Rogers.

1925   *Baltimore Sun* journalist H. L. Mencken (1880–1956) met with famed attorney Clarence Darrow on May 14 to urge him to defend John Scopes. Mencken—who coined the phrases *Bible Belt* and *Monkey Trial*—covered the trial in Dayton (where he was described as "the most respected, hated, reviled, feared, and loved person" in Tennessee) and wrote 13 articles for the *Sun*, work that is regarded as some of the greatest journalism in American history. In the most famous of his articles, Mencken described Scopes' trial as a "religious orgy." Mencken later described fundamentalists as being "everywhere

where learning is too heavy a burden for mortal minds to carry." Mencken shaped, as well as reported, the trial, and even Scopes admitted that the trial "was Mencken's show," and that "a mention of the Dayton trial more likely invokes Mencken than it does me." Although William Jennings Bryan once described Mencken as "the best newspaperman in the country," Mencken despised Bryan, and his first comment upon hearing of Bryan's death was "We killed the son-of-a bitch." Bryan's death did not slow Mencken's attack; indeed, Mencken told readers that if Bryan was sincere, "then so was P. T. Barnum." Mencken's "In Memoriam: W. J. B.," a masterpiece of invective, was taken at face value by Jerome Lawrence and Robert Lee when creating the Bryan-esque character Matthew Brady for their influential play *Inherit the Wind*. Just before the start of the Scopes trial, Mencken marked the one-hundredth anniversary of Thomas Huxley's birth by praising him as "the greatest Englishman of the Nineteenth Century—perhaps the greatest Englishman of all time."

1925   As the Scopes trial approached, the American Telephone & Telegraph Company installed 10.5 miles of temporary lines to speed the transmission of stories (*New York Times* reporters alone telegraphed more than 100,000 words about the trial). On the first day of the trial, Vernon Dalhart (born Marion Try Slaughter [1883–1948])—a popular country singer—popularized the trial by recording Carson Robison's (1890–1957) *The John T. Scopes Trial* for the Columbia Phonograph Company (#15037-D). Weeks later, Dalhart also recorded *Bryan's Last Fight* (Columbia Records #15039), which proclaimed that Bryan "stood for his own convictions, and for them he'd always fight."

1925   Almost 2,000 spectators watched the Scopes trial reach its climax on the lawn of the courthouse: after earlier baiting Bryan by saying that "Bryan has not dared test his views in open court under oath," Defense lawyer Arthur Garfield Hays announced to the court, "The defense desires to call Mr. Bryan as a witness." Bryan did not have to testify (Judge Raulston left the decision to Bryan), but Bryan—falling into Darrow's trap—took the witness stand. In the 90-minute examination, Darrow referred to Bryan's "fool religion" and questioned Bryan about his "fool ideas" (e.g., Jonah being swallowed by a whale, Joshua's commanding the sun to stand still to lengthen the day). Bryan, who was less concerned about Earth's age than the influence of evolution on societal morals, eventually admitted that he did not believe in a literal interpretation of the Bible, and instead endorsed day-age creationism. *The New York Times* described the Darrow-Bryan encounter as "an absurdly pathetic performance," and noted, "Darrow succeeded in showing that Bryan knows little about the science of the world." Throughout Scopes's trial, John Straton provided a daily editorial for the Hearst newspapers.

1927   Sinclair Lewis (1885–1951) wrote *Elmer Gantry*, a best-selling novel. The main character in the satirical story—a hypocritical evangelist—was based partly on antievolution crusader John Roach Straton. Lewis dedicated the book to H. L. Mencken "with profound admiration."

1931   Frederick Allen's (1890–1954) influential history *Only Yesterday* portrayed the Scopes trial as blind fundamentalism versus enlightened skepticism, and became the standard interpretation of events in Dayton. This interpretation was later reinforced by *Inherit the Wind*.

1936   English ornithologist Percy Lowe's (1870–1948) paper "The Finches of the Galapagos in Relation to Darwin's Conception of Species" in the journal *Ibis* introduced the term *Darwin's finches*. However, the term was not made famous until the appearance of David Lack's 1947 book *Darwin's Finches*.

1940   Will Houghton (1887–1947), president of the Moody Bible Institute, argued in *Moody Monthly* that Nazism was based on evolution and materialism, and that universities were responsible for modernism having replaced orthodox faith with Darwinism and Marxism.

1946   Seventh-Day Adventist Harold Clark's geology textbook *The New Diluvialism* claimed that rising floodwaters successively destroyed ecological zones, thereby producing the predictable arrangement of fossils visible today. Unlike George Price, Clark conceded the validity of the geologic column. Clark's so-called "ecological zonation theory," which remains popular among some creationists, enabled creationists to accept the validity of the geological column while rejecting an ancient

Earth. However, Clark's idea was rejected by scientists for many reasons, including the fact that some organisms (e.g., corals and clams) appear in virtually all strata, and that organisms living in the same ecological zone (e.g., fish and whales, birds, and flying reptiles) appear in different strata. Clark's idea was also distasteful to George Price; *The New Diluvialism* prompted Price to challenge Clark with *Theories of Satanic Origin*.

1951   British physician and geneticist Henry Bernard Davis (H. B. D.) Kettlewell (1907–1979) begins studying the peppered moth, *Biston betularia*. Kettlewell's lab work showed that dark-winged moths tended to choose dark backgrounds, and light-winged moths tended to choose lighter backgrounds. Field experiments demonstrated that birds could function as selective agents, and that birds had to learn to recognize a type of prey before they could exploit it. Kettlewell confirmed his claims with mark-release-recapture experiments in polluted forests near Birmingham and in pristine forests near Dorset. Kettlewell's work showed that lightly colored moths were more conspicuous than darkly colored moths in polluted areas, and in those areas were more susceptible to predation by birds. Nobel laureate Niko Tinbergen's (1907–1988) movies of the differing predation rates of the moths by birds were shown at science meetings throughout the world, and Kettlewell's work appeared in a *Scientific American* article titled "Darwin's Missing Evidence." Kettlewell's work was described in 1978 by Sewall Wright as "the clearest case in which a conspicuous evolutionary process has actually been observed," and for many years was cited in biology textbooks as the best example of natural selection in action.

1955   The play *Inherit the Wind* opens to favorable reviews in Dallas, Texas. Three months later it opened at Broadway's National Theatre for a three-year run. The play—a response to the threat to intellectual freedom posed by the anti-Communist hysteria of the McCarthy era (then safely a generation in the past)—described "the famous 'Monkey Trial' that rocked America" as "the most explosive trial of the century." When its Broadway run ended, *Inherit the Wind* was the most successful and longest-running drama in Broadway's history. In 1957, the success of the play prompted the *Encyclopedia*

Britannica to include the Scopes trial. The movie version of *Inherit the Wind* in 1960 strongly shaped the public's view of the Scopes trial and the evolution-creationism controversy.

1960   *The Flintstones*, television's first prime-time animated series, debuted on ABC. The show, which portrayed a "modern stone-age family" of humans who lived with dinosaurs, saber-toothed tigers, wooly mammoths, and other extinct animals, lasted 166 episodes and indoctrinated millions of children and adults with the notion that humans lived contemporaneously with dinosaurs. This claim was repeatedly rejected by biologists, geologists, and other scientists, but has remained a foundation of young-Earth creationism (e.g., that dinosaurs and humans were created on the sixth day of creation). Indeed, the Creation Museum, operated by the antievolution organization Answers in Genesis (AIG), includes dioramas depicting humans living with dinosaurs. In 2002, NSF reported in its *Science and Technology Indicators* that 48 percent of Americans believed that humans and dinosaurs lived at the same time.

1963   The BSCS published its three biology textbooks, each of which emphasized evolution. The books were described not by their foci (e.g., ecology, molecular biology, cell biology), but instead by their colors (i.e., green, blue, yellow) to avoid the implications that the books were specialized for advanced biology courses or that BSCS was trying to establish a national curriculum for biology. Some states accepted the BSCS textbooks, but education officials in Texas insisted that publishers delete statements such as, "To biologists there is no longer any reasonable doubt that evolution occurs." Publishers accepted these revisions, which appeared in subsequent editions of the blue version. Within a few years after being released, BSCS books were being used in nearly half of all high school biology courses in the United States. They were also used in several other countries; Ken Ham, the founder of the antievolution organization AIG, taught from a BSCS textbook when he was a public school teacher in Australia in 1975. Publishers of competing books soon began producing similar books.

1963   BSCS published *Biology Teachers Handbook*, which proclaimed, "It is no longer possible to give a complete or even a coherent account of living things without the story of evolution."

1963   Claiming that BSCS's textbooks are the "most vicious attack
we have ever seen on the Christian religion," Church of
Christ preacher Reuel Gordon Lemmons (1912–1989) of
Austin, Texas, campaigned to block the adoption of pro-
evolution textbooks in Texas. Like Nell Segraves and Jean
Sumrall in California, Lemmons demanded that evolution be
taught "as a theory" and was helped by antievolution groups
(e.g., the CRS) and Mel and Norma Gabler. The Gablers
believed that the humanistic emphasis of the BSCS textbooks
would lead to the rejection of traditional morality.

1964   In *Essays of a Humanist*, Julian Huxley claimed that "God is a
hypothesis constructed by man to help him understand what existence is all about" and that "evolution is a process, of which we are
products, and in which we are active agents. There is no finality
about the process, and no automatic or unified progress. . . ."

1965   NBC produced and broadcast a new version of *Inherit the
Wind*. Meanwhile, John Scopes claimed "restrictive legislation
on academic freedom is forever a thing of the past."

1979   In the popular television broadcast of *In Search of Noah's Flood*,
actor Leonard Nimoy (b. 1931) noted that "no written word
has survived as much skepticism as the story of Noah's Ark."

1981   After an Associated Press-NBC News poll showed that almost
75 percent of parents and teachers accepted the teaching of creationism in public schools, the Tampa, Florida, school board
required equal time for creationism and evolution. This decision, opposed by many science teachers, made the study of creationism mandatory for its 115,000 students. Soon thereafter, a
debate in Tampa involving the Institute for Creation
Research's Henry Morris was attended by 1,700 spectators
and covered by seven radio stations, six television stations,
and numerous newspapers.

1991   Phillip Johnson wrote *Darwin on Trial*, a critique of Darwinian
evolution. The book proved to be a key text for the intelligent
design movement.

1995   The Discovery Institute of Seattle formed the Center for the
Renewal of Science and Culture, later renamed the Center for
Science and Culture, with a goal of promoting the intelligent
design movement.

1998   In the popular *Darwin's Leap of Faith: Exposing the False Religion of Evolution*, television personalities John Ankerberg (b. 1945) and John Weldon argued that Darwin never considered his theory of evolution to be overly convincing and that he devised it because it was "convenient to his rejection of God."

1999   Chinese paleontologist Xiao-Chun Wu and his colleagues discover *Sinornithocaurus*, a meat-eating dinosaur, in volcanic ash deposited 120 million years ago in China. *Sinornithocaurus* had a coat of "proto-feathers" conceivably used for insulation, but not for flight. This discovery supported the claims that birds evolved from dinosaurs. The same year, a fossil appeared from China's Liaoning Province that seemed to be a feathered bird with a long dinosaur-like tail. The fossil, bought at a gem show in Arizona, was described by the press as a "missing link" and named *Archaeoraptor* in an article published in *National Geographic*. Scientists later showed the fossil to be a fake; the head and upper body were from a primitive fossil bird (*Yanomis*) and the tail from *Microraptor*, a small, gliding theropod. Although scientists have found many examples of feathered dinosaurs, creationists continue to use the *Archaeoraptor* scandal to try to question evolutionary theory.

1999   After fatal shootings at Colorado's Columbine High School, Texas Congressman Tom DeLay (b. 1947) blamed the tragedy on "school systems [that] teach the children that they are nothing but glorified apes who evolutionized out of some primordial soup of mud." Although this quote was attributed to DeLay, he read it in an article by Addison Dawson that had recently appeared in the *San Angelo Standard-Times*.

1999   The Republican-dominated Kansas board of education endorsed, by a vote of 6–4, a set of science education standards for the state's 305 public school districts. These standards—developed by the Creation Science Association for Mid-America—included no mention of biological macroevolution, the age of the Earth, or the origin and early development of the universe. The board had earlier rejected standards developed by a twenty-seven-member panel of scientists and science educators. *Science* reported the story as "Kansas Dumps Darwin," and *The New York Times* described the vote with the headline "Board for Kansas Deletes Evolution from Curriculum—A Creationist Victory." The AAAS described the decision as "a serious disservice to students and teachers." Soon after the Kansas decision, the Kentucky Department of Education deleted the word *evolution* from its educational guidelines, and conservative columnist and political commentator George Will (b. 1941) noted, "Every [political] party at any given time has a certain set of issues on its fringe that can make it look strange, and this is one that can make the Republicans look strange." An editorial in *Scientific American* suggested that college admissions boards question the qualifications of applicants from Kansas. However, antievolution crusader Phillip Johnson (who had given money and public support to a creationist school board candidate) called the board's decision "courageous," while biochemist and ID-advocate Michael Behe called it "heartening."

1999   *Time* magazine published an article titled "Up from the Apes" that began by stating, "Despite the protests of creationists . . . science has long taught that human beings are just another kind of animal. . . . " Impassioned creationists denounced the article as propaganda.

1999   In an interview with the *Sunday Telegraph*, Richard Dawkins argued that "evolution should be one of the first things you learn at school . . . and what do [children] get instead? Sacred hearts and incense. Shallow, empty religion."

2000   The Ottawa (Canada) *Citizen* reported that a new curriculum designed to avoid controversy would ensure that most students in Ontario would go through elementary and high school without being taught evolution.

2000   The Showtime Network produced and broadcast another version of *Inherit the Wind*.

2001   The Public Broadcasting Service aired the seven-part, eight-hour series *Evolution*, which, according to *The Christian Science Monitor*, "[did] its best not only to explain Charles Darwin's theory of the origins of material life, but to take seriously conservative Christians' religious objections to it." The series received a wide viewership and acclaim from scientists and the popular press, but was not without its critics. For example, in response to pressure from State Senator Stan Hawkins, Idaho Public Television aired "The Young Age of the Earth" (produced by Earth Science Associates of Knoxville, Tennessee). The Discovery Institute released the 154-page "Getting the Facts Straight: A Viewer's Guide to PBS's Evolution," which claimed that "[b]y systematically ignoring the bigger picture, *Evolution* distorts the issues and misleads its viewers." To counter classroom guides posted on PBS's web site, Gary Luskin of the Intelligent Design and Evolution Awareness (IDEA) Center released the antagonistic "Ten Questions to Ask Your Students about the PBS Evolution Series," and in an article titled "911 Rang Again," Ken Cumming of the Institute for Creation Research compared the terrorist bombings of September 11, 2001, to the airing of the *Evolution* series, saying "while the public now understands from President Bush that 'We're at War' with religious fanatics around the world, they don't have a clue that America is being attacked from within through its public schools by a militant religious movement called Darwinists."

2001   The Discovery Institute launched the web site *A Scientific Dissent from Darwinism*, which asked those holding a Ph.D. in science, engineering, mathematics, or computer science to endorse the statement, "We are skeptical of claims for the ability of random mutation and natural selection to account for the complexity of life." (As of June 2012, more than 800 signatures had been collected.) At the same time, the Discovery Institute published full-page advertisements in three well-known national periodicals with the same headline, and as about 100 scientists who endorsed the claim as well. The NCSE responded with its *Project Steve* parody, which asked Ph.D.-level scientists named "Steve"—which should be about 1 percent of all scientists, and which was chosen in honor of the late Stephen Gould—to sign a letter supporting evolution. As of June 2012, more than 1,200 names had been collected.

2004   Michael Behe and David Snoke published a paper in the peer-reviewed journal *Protein Science* noting, "Although many scientists assume that Darwinian processes account for the evolution of complex biochemical systems, we are skeptical."

2004   In an article titled "OK, We Give Up" in its April Fool issue, *Scientific American* apologized for promoting evolution by noting that: "As editors, we had no business being persuaded by mountains of evidence ... [or] thinking that scientists understand their fields better than, say, U.S. senators or best-selling novelists do."

2004   The anti-ID film *Flock of Dodos* received its first public screening in Kansas, a state again in the midst of a controversy over the teaching of evolution in its public schools. Randy Olson, a marine biologist, made *Flock of Dodos* to examine the claims of the ID movement, as well as the relatively weak response the scientific community had historically mounted to challenges to its authority.

2005   In an article titled "Twilight for the Enlightenment?," the editor of *Science* likened the widespread rejection of evolution in the United States to pre-Enlightenment adherence to superstition and irrational beliefs about the world.

2005   An effort to require the teaching of intelligent design in the public schools prompted a U.S. District Court trial in Dover, Delaware. The judge's decision in *Kitzmiller v. Dover Area School District* finds that the intelligent design movement has its roots in Christianity and is therefore a violation of the First Amendment's prohibition against the government establishing a religion.

2005   *Time* used its cover to promote the "Evolution Wars" and the growing controversy about teaching ID. The article drew hundreds of impassioned responses. Meanwhile, in *Newsweek*, conservative columnist George Will noted, "The problem with intelligent-design theory is not that it is false but that it is not falsifiable. Not being susceptible to contradicting evidence, it is not a testable hypothesis. Hence it is not a scientific but a creedal tenet—a matter of faith, unsuited to a public school's science curriculum."

2006   Famed biologist E. O. Wilson, writing in *USA Today*, claimed that the disagreements between science and fundamentalist Christianity are "unsoluble. ... The two world views— science-based explanations and faith-based religions—cannot be reconciled."

2007   PBS aired *Judgment Day: Intelligent Design on Trial*, a documentary recounting the 2005 *Kitzmiller v. Dover Area Independent*

School District trial on the issue of teaching intelligent design in the public schools.

2008 *Expelled: No Intelligence Allowed*, a documentary promoted by the Discovery Institute, opened nationwide, and promising to "expose the frightening agenda of the 'Darwinian Machine.' " The movie claimed that conspiring scientists suppressed criticisms of evolution and that evolution was responsible for societal ills such as the Nazi Holocaust. The film was a critical failure that received poor reviews; for example, *The New York Times* described it as "a conspiracy-theory rant masquerading as investigative inquiry ... [,] an unprincipled propaganda piece that insults believers and nonbelievers alike." It was also a box-office flop; the movie grossed only $7.9 million and was out of theatres in only eight weeks. *Expelled*'s star, comedian Ben Stein, dismissed critics as "the self-appointed atheist elite."

2009 The bicentennial of Charles Darwin's birth produced a flood of books, exhibits, symposia, merchandise, collectibles (including a £2 coin), and celebrations honoring Darwin's impact on science and society. February editions of scientific journals (e.g., *Nature, Science*) and popular magazines (e.g., *National Geographic, Scientific American*) selected cover art to commemorate the bicentennial of Charles Darwin's birth. Famed biologist E. O. Wilson remarked, "Charles Darwin's *On the Origin of Species* can fairly be ranked as the most important book ever written."

2011 NASA's Jet Propulsion Laboratory in California demoted and then fired computer specialist David Coppedge. He then filed a lawsuit that contended NASA retaliated against him because he supports intelligent design and discuss ID with his coworkers.

2012 The popular movie *Prometheus*, by director Ridley Scott, depicted an alien life form creating life on Earth. Some intelligent design proponents as a subtle argument supporting intelligent design, although many movie reviewers dispute that interpretation.

—Randy Moore, Mark Decker, and Sehoya Cotner

Resources and References

SELECTED ANNOTATED SOURCES

Beckwith, Francis J. *Law, Darwinism, and Public Education: The Establishment Clause and the Challenge of Intelligent Design*. Lanham, MD: Rowan and Littlefield, 2003. This book is a pro-intelligent design work based on the idea that teaching intelligent design in public school science classes does not violate the First Amendment's prohibition against establishing an official government religion.

Behe, Michael J. *Darwin's Black Box: The Biochemical Challenge to Evolution*. New York: Free Press, 1996. Behe's book develops the idea of irreducible complexity of some biological systems and argues for intelligent design.

Bowler, Peter J. *Monkey Trials and Gorilla Sermons: Evolution and Christianity from Darwin to Intelligent Design*. Cambridge, MA: Harvard University Press, 2007. This book is an account of the history of evolution and criticism of it, including by intelligent design proponents.

Browne, Janet. *Darwin's Origin of Species: A Biography*. New York: Atlantic Monthly Press, 2006. Browne provides a brief history of how Darwin came to write *The Origin of Species*, as well as the aftermath of Darwin's book's publication.

Buddenbaum, Judith M. *Reporting News about Religion*. Ames, IA: Iowa State University Press, 1998. Buddenbaum examines the ways the news media report about various religious issues, including a section on how the media frame news about religion.

Campbell, John Angus, and Steven C. Meyer, eds. *Darwinism, Design, and Public Education*. East Lansing, MI: Michigan State University Press,

2003. This edited collection provides an even-handed approach, with some chapters advocating intelligent design and others arguing on behalf of evolution.

Caudill, Edward. *Darwinism in the Press: The Evolution of an Idea*. Hillsdale, NJ: Lawrence Erlbaum Associates, 1989. Caudill, a University of Tennessee communications professor, provides as account of how Darwin's theory was reported by the press, as well as how the theory has been misused to promote social issues.

Caudill, Edward. *Darwinian Myths: The Legends and Misuses of a Theory*. Knoxville: University of Tennessee Press, 1997. Caudill looks at how Darwin's theory was adapted and misused by others, including the application of Darwin's theory to social issues.

Darwin, Charles. *On the Origin of Species: A Facsimile*. Cambridge, MA: Harvard University Press, 1966. This is a modern edition of Darwin's original text.

Davis, Percival, and Dean H. Kenyon. *Of Pandas and People: The Central Question of Biological Origins*. 2nd ed. Dallas: Haughton Publishing, 1993. A pro-intelligent design textbook aimed at middle school and high school science classes.

Dembski, William A., and Michael Ruse, eds. *Debating Design: From Darwin to DNA*. Cambridge: Cambridge University Press, 2004. This collection of essays includes arguments before and against intelligent design, with some of the leading figures on both sides of the issue included.

Dembski, William A., and JonathanWitt. *Intelligent Design Uncensored*. Downers Grove, IL: IVP Books, 2010. Written in a simple style suitable for high school or college students, Dembski and Witt promote intelligent design as a legitimate science.

Dembski, William A. *The Design Revolution: Answering the Toughest Questions about Intelligent Design*. Downers Grove, IL: InterVarsity Press, 2004. Dembski's book is one of the major works detailing the intelligent design movement and arguing that it is a science as legitimate as the study of evolution.

Dembski, William, and Sean McDonald. *Understanding Intelligent Design*. Eugene, OR: Harvest House Publishers, 2008. This is a popularly written explanation of intelligent design by Dembski, one of the leaders of the movement, and McDonald, a theologian and Bible professor.

Foster, John Bellamy, Brett Clark, and Richard York. *Critique of Intelligent Design: Materialism versus Creationism from Antiquity to the Present*. New York: Monthly Review Press, 2008. This book presents a critique of both the intelligent design movement and creationism.

Gould, Stephen Jay. *The Panda's Thumb: More Reflections in Natural History*. New York: W.W. Norton, 1980. Gould provides a colorfully written defense of evolution.

House, H. Wayne, ed. *Intelligent Design 101*. Grand Rapids, MI: Kregel Publications, 2008. This edited collection includes chapters by the leading figures in the intelligent design movement, including Phillip Johnson, William Dembski, and Michael Behe.

Howard, Jonathan. *Darwin: A Very Short Introduction.* Oxford: Oxford University Press, 2001. As the title implies, this is a brief look at Darwin's life.

Hume, Edward. *Monkey Girl: Evolution, Education, Religion, and the Battle for America's Soul.* New York: HarperCollins, 2007. A decidedly pro-evolution, anti-intelligent account of the Dover, Pennsylvania, legal case over teaching intelligent design in the public schools.

Johnson, Phillip. *Darwin on Trial.* Downers Grove, IL: InterVarsity Press, 1991. Written as a legal brief questioning the validity of evolution, Johnson's book is credited by many with starting the intelligent design movement.

Luskin, Casey. "It's Constitutional but Not Smart to Teach Intelligent Design in Schools." http://www.beliefnet.com/News/2005/11/Its-Constitutional -But-Not-Smart-To-Teach-Intelligent-Design-In-Schools.asp. Luskin is one of the leaders of the intelligent design movement, and this work argues why teaching about intelligent design should not be required in public schools.

Martin, Justin D. *Religion, Science and Public Education: Newspaper Coverage of the Origins' Debate in Ohio Public Schools.* Master's thesis, University of Florida, 2004. Martin's thesis examines newspaper framing of the intelligent design-evolution debate during the fight over the Ohio Board of Education's decision on whether to require public school science teachers to include intelligent design in their curriculum.

Miller, Kenneth R. *Only a Theory: Evolution and the Battle for America's Soul.* New York: Penguin, 2008. A leading advocate of evolution dissects what he says are the intelligent design movement's beliefs.

Neuman, W. Russell, Marion R. Just, and Ann N. Crigler. *Common Knowledge: News and the Construction of Political Meaning.* Chicago: University of Chicago Press, 1992. This book is an explanation of how media framing works, including an extensive discussion of the theory behind media framing analysis.

Petto, Andrew J., and Laurie R. Godfrey, eds. *Scientists Confront Intelligent Design and Creationism.* New York: W.W. Norton, 2007. This collection of chapters by leading evolution proponents compares intelligent design to creationism and critiques the intelligent design movement.

Poole, Steven. *Unspeak.* New York: Grove Press, 2006. This book examines how public discussion of issues helps to drive changes in public opinion. It includes a chapter on how the intelligent design movement has been framed.

Quammen, David. *The Reluctant Mr. Darwin: An Intimate Portrait of Charles Darwin and the Making of His Theory of Evolution.* New York: W.W. Norton, 2006. This is another brief biography of Charles Darwin and how he developed the theory of evolution.

Reese, Stephen D., Oscar H. Gandy, and August E. Grant, eds. *Framing Public Life*. Mahwah, NJ: Lawrence Erlbaum Associates, 2001. This edited collection by media researchers provides several examples of how media framing works, including several studies of framing particular issues.

Ruse, Michael. *Darwin and Design: Does Evolution Have a Purpose?* Cambridge, MA: Harvard University Press, 2003. This presents a historical account of the development of evolution and opposition to it from intelligent design advocates.

Ruse, Michael. *Darwin and Design: Does Evolution Have a Purpose?* Cambridge, MA: Harvard University Press, 2003. This book is a simply written biography of Charles Darwin, with plenty of quotes from Darwin's books and other contemporary works, focusing on the philosophical underpinnings of evolution.

Scott, Eugenie C. *Evolution vs. Creationism: An Introduction*. 2nd ed. Westport, CT: Greenwood Press, 2009. An even-handed account of the evolution versus creationism/intelligent design movement from Scott, a leading proponent of teaching evolution in the schools.

Scott, Eugenie C., and Glenn Branch, eds. *Not in Our Classroom: Why Intelligent Design Is Wrong for Our Schools*. Boston: Beacon Press, 2006. This edited collection discusses why the authors believe intelligent design does not belong in public school classrooms.

Wells, Jonathan. *Icons of Evolution: Science or Myth*. Washington, DC: Regnery Publishing, 2000. This book criticizes some of the "icons" of the evolution movement, including the concept of the evolution of humans from apes. It is considered to be one of the leading books criticizing evolution.

www.discovery.org. The main web site for the conservative Discovery Institute, a pro-intelligent design organization. The link to the Center for Science and Culture, its main intelligent design division, is www.discovery.org.csc/org.

www.law.umkc.edu/faculty/projects/ftrials/scopes/scopes.htm. This extensive web site, at the University of Missouri-Kansas City Law School, provides exhaustive material about John Scopes and the "monkey" trial in Dayton, Tennessee, including archival photographs and original material, and even a few short video clips from the 1925 trial.

REFERENCES

"About the Museum." The Creation Museum. Accessed September 6, 2010. http://creationmuseum.org/about.

Applebome, Peter. "Pope Shows How Faith and Evolution Coexist." *New York Times*, October 25, 1996, A12.

Andsager, Julie. L. "How Interest Groups Attempt to Shape Public Opinion with Competing News Frames." *Journalism and Mass Communication Quarterly* 77 no. 3 (2000), 577–592.

Badkhen, Anne. "Anti-Evolution Teachings Gain Foothold in U.S. Schools." *San Francisco Chronicle*, November 30, 2004, A1.

Beckwith, Francis J. *Law, Darwinism, and Public Education: The Establishment Clause and the Challenge of Intelligent Design*. Lanham, MD: Rowan and Littlefield, 2003.

Behe, Michael J. "Irreducible Complexity: Obstacle to Darwinian Evolution." In *Debating Design: From Darwin to DNA*, edited by William A. Dembski and Michael Ruse, 352–370. Cambridge: Cambridge University Press, 2004.

Behe, Michael J. *Darwin's Black Box: The Biochemical Challenge to Evolution*. New York: Free Press, 1996.

Boffey, Philip M. "100 Years after Darwin's Death, His Theory Still Evolves." *New York Times*, April 20, 1982, C2.

Bowler, Peter J. *Evolution: The History of an Idea*. Berkeley: University of California Press, 1989.

Bowler, Peter J. *Monkey Trials and Gorilla Sermons: Evolution and Christianity from Darwin to Intelligent Design*. Cambridge, MA: Harvard University Press, 2007.

Browne, Janet. *Darwin's Origin of Species: A Biography*. New York: Atlantic Monthly Press, 2006.

Buddenbaum, Judith M. *Reporting News about Religion*. Ames, IA: Iowa State University Press, 1998.

Caudill, Edward. *Darwinian Myths: The Legends and Misuses of a Theory*. Knoxville: University of Tennessee Press, 1997.

Caudill, Edward. *Darwinism in the Press: The Evolution of an Idea*. Hillsdale, NJ: Lawrence Erlbaum Associates, 1989.

Caudill, Edward. "Intelligently Designed: Creationism's News Appeal." *Journalism and Mass Communication Quarterly* 87, no. 1 (Spring 2010): 84–99.

Cavanagh, Sean. "Evolution Loses and Wins, All in One Day." *Education Week* 25, no. 12 (November 18, 2005): 1.

CNN. "Transcripts." August 23, 2005. Accessed October 5, 2010. http://transcripts.cnn.com/TRANSCRIPTS/0508/23/lkl.01.html.

Coyne, Jerry A. "Intelligent Design: The Faith That Dare Not Speak Its Name." In *Intelligent Thought: Science versus the Intelligent Design Movement*, edited by John Bockman, 3–23. New York: Vintage, 2006.

Darwin, Charles. *On the Origin of Species: A Facsimile*. Cambridge, MA: Harvard University Press, 1966.

Davis, Percival, and Dean H. Kenyon. *Of Pandas and People: The Central Question of Biological Origins*. 2nd ed. Dallas: Haughton Publishing, 1993.

Davis, Percival, Dean H. Kenyon, and Charles B. Thaxton. *Of Pandas and People: The Central Question of Biological Origins*. Dallas: Haughton Publishing, 1989.

"Defense Gets Its Days in Court in Support of Intelligent Design." *Education Week*, October 26, 2005: 6.

Dembski, William A. *The Design Inference: Eliminating Chance through Small Probabilities*. Cambridge: Cambridge University Press, 1998.

Dembski, William A. *The Design Revolution: Answering the Toughest Questions about Intelligent Design.* Downers Grove, IL: InterVarsity Press, 2004.

Dembski, William A. "The Logical Underpinnings of Intelligent Design." In *Debating Design: From Darwin to DNA,* edited by William A. Dembski and Michael Ruse, 311–330. Cambridge: Cambridge University Press, 2004.

Dembski, William A., and Jonathan Witt. *Intelligent Design Uncensored.* Downers Grove, IL: IVP Books, 2010.

Dembski, William A., and Michael Ruse. "General Introduction." In *Debating Design: From Darwin to DNA,* edited by William A. Dembski and Michael Ruse, 3–12. Cambridge: Cambridge University Press, 2004.

Dembski, William, and Sean McDonald. *Understanding Intelligent Design.* Eugene, OR: Harvest House Publishers, 2008.

Dennett, Daniel C. "Show Me the Science." *New York Times,* August 28, 2005.

Dennett, Daniel C. "The Hoax of Intelligent Design and How It Was Perpetrated." In *Intelligent Thought: Science Versus the Intelligent Design Movement,* edited by John Brookman, 33–49. New York: Vintage Books, 2006.

Denton, Michael. *Evolution: A Theory in Crisis.* Bethesda, MD: Adler and Adler, 1985.

Deslatte, Melinda. "No Religion Allowed in La. Science Classes." *Associated Press,* January 15, 2009.

DeWolf, David K. "The 'Teach the Controversy' Controversy." *University of St. Thomas Journal of Law & Public Policy* 4, no. 1 (Fall 2009): 326–353.

DeWolf, David K., Stephen C. Meyer, and Mark E. DeForest. "Teaching the Controversy: Is It Science, Religion, or Speech?" In *Darwinism, Design, and Public Education,* edited by John Angus Campbell and Steven C. Meyer, 59–132. East Lansing, MI: Michigan State University Press, 2003.

DeWolf, David K., John G. West, and Casey Luskin. "Intelligent Design Will Survive *Kitzmiller v. Dover.*" *Montana Law Review* 68, no. 7 (2007): 7–57.

Discovery Institute. "About Discovery." Accessed August 20, 2010. http://www.discovery.org/about.php.

Discovery Institute. "Darwinian Evolution, Intelligent Design and Education Policy." Accessed August 19, 2010. http://www.intelligentdesign.org/education.php.

Discovery Institute. "Definition of Intelligent Design." Accessed September 3, 2010. http://www.intelligentdesign.org/whatisid.php.

Discovery Institute. "Join the Free Speech on Evolution Campaign." Accessed August 19, 2010. http://www.discovery.org/csc/freeSpeechEvol CampMain.php.

Discovery Institute. "Media Backgrounder: Intelligent Design Article Sparks Controversy." September 7, 2004. Accessed August 20, 2010. http://www.discovery.org/a/2190.

Discovery Institute. "The 'Wedge Document': So What?" February 3, 2006. Accessed August 20, 2010. http://www.discovery.org/a/2101.

Dodson, Edward O., and Peter Dodson. *Evolution: Process and Product.* Belmont, CA: Wadsworth Publishing Co, 1985.

Entman, Robert M. "Framing: Toward Clarification of a Fractured Paradigm." *Journal of Communication* 43, no. 4 (1993): 51–58.

Fisher, Ian. "About Creation, Pope Melds Faith with Science." *New York Times,* April 12, 2007, A6.

Flam, Fay. "O'Donnell's Misperception Is Common." *Philadelphia Inquirer,* October 4, 2010, D1.

Foster, John Bellamy, Brett Clark, and Richard York. *Critique of Intelligent Design: Materialism versus Creationism from Antiquity to the Present.* New York: Monthly Review Press, 2008.

Fox News. "O'Reilly v. Atheist Author Richard Dawkins." October 12, 2009. Accessed October12, 2010. http://www.foxnews.com/story/0,2933 ,564422,00.html.

Frame, John M. "Is Intelligent Design Science?" Accessed September 10, 2010. http://www.frame-poythress.org/frame_articles/IntelligentDesign .htm.

Gamson, William A. "News as Framing." *American Behavioral Scientist* 33, no. 2 (1989), 157–161.

Gamson, William A., and Andre Modigliani. "Discourse and Public Opinion on Nuclear Power: A Constructionist Approach." *American Journal of Sociology* 95, no. 1 (1989): 1–37.

Gitlin, Todd. *The Whole World Is Watching: Mass Media in the Making and Unmaking of the New Left.* Berkeley: University of California Press, 1980.

Goldstein, Laurie. "Closing Arguments Made in Trial on Intelligent Design." *New York Times,* November 5, 2005, A14.

Gould, Stephen Jay. *Ever Since Darwin: Reflections in Natural History.* New York: W.W. Norton, 1977.

Gould, Stephen Jay. *The Panda's Thumb: More Reflections in Natural History.* New York: W.W. Norton, 1980.

Grimm, Joshua. " 'Teach the Controversy': The Relationship between Sources and Frames in Reporting the Intelligent Design Debate." *Science Communication* 31, no. 2 (2010): 167–186.

Guervara, Emily. "State Board of Education Approves Evolution Debate." *Beaumont (TX) Enterprise,* March 27, 2009.

Gunn, Angus M. *Evolution and Creationism in Public Schools.* Jefferson, NC: McFarland, 2004.

Harrower, Tim. *Inside Reporting: A Practical Guide to the Craft of Journalism.* New York: McGraw-Hill, 2007.

Hedlun, Patric. "False Information on Intelligent Design Course Given to Board." *Mountain (Frazier Park, CA) Enterprise,* December 30, 2005.

Hoover, Stewart M. *Religion in the News: Faith and Journalism in American Public Discourse.* Thousand Oaks, CA: SAGE Publications, 1998.

Howard, Jonathan. *Darwin: A Very Short Introduction.* Oxford: Oxford University Press, 2001.

Hoyle, F., and N. C. Wickramasinghe. *Evolution from Space: A Theory of Cosmic Creationism*. New York: Simon and Schuster, 1982.

Hume, Edward. *Monkey Girl: Evolution, Education, Religion, and the Battle for America's Soul*. New York: HarperCollins, 2007.

Johnson, Phillip. "Bringing Balance to a Fiery Debate." In *Intelligent Design 101*, edited by H. Wayne House, 21–40. Grand Rapids, MI: Kregel Publications, 2008.

Johnson, Phillip. *Darwin on Trial*. Downers Grove, IL: InterVarsity Press, 1991.

Johnson, Vicki D. "A Contemporary Controversy in American Education: Including Intelligent Design in the Science Curriculum." *The Educational Forum* 70 (Spring 2006): 222–236.

Kaufman, Marc. "Modern Man, Neanderthals Seen as Kindred Spirits." *Washington Post*, April 30, 2007, A6.

Kuhn, Thomas. *The Structure of Scientific Revolutions*. Chicago: University of Chicago Press, 1962.

Lester, Will. "64% Want Creationism Taught with Evolution." *Chicago Sun-Times*, September 1, 2005.

Luskin, Casey. "A Brief History of Intelligent Design." The Discovery Institute. September 8, 2006. Accessed August 20, 2010. http://www.discovery.org/a/8931.

Luskin, Casey. "Finding Intelligent Design in Nature." In *Intelligent Design 101*, edited by H. Wayne House, 69–111. Grand Rapids, MI: Kregel Publications, 2008.

Luskin, Casey. "It's Constitutional but Not Smart to Teach Intelligent Design in Schools." Accessed September 14, 2010. http://www.beliefnet.com/News/2005/11/Its-Constitutional-But-Not-Smart-To-Teach-Intelligent-Design-In-Schools.aspx.

Martin, Justin D. *Religion, Science and Public Education: Newspaper Coverage of the Origins' Debate in Ohio Public Schools*. Master's thesis, University of Florida, 2004.

Martin, Justin, D., Kaye D. Trammell, Daphne Landers, Jeanne M. Valois, and Terry Bailey. "Journalism and the Debate over Origins: Newspaper Coverage of Intelligent Design." *Journal of Religion and Media* 5, no. 1 (2006): 49–61.

Matzke, Nicholas J., and Paul R. Gross. "Analyzing Critical Analysis: The Fallback Antievolutionary Strategy." In *Not in Our Classroom: Why Intelligent Design Is Wrong for Our Schools*, edited by Eugenie C. Scott and Glenn Branch, 28–56. Boston: Beacon Press, 2006.

McCune, Cynthia A. "Framing Reality: Shaping the News Coverage of the 1996 Tennessee Debate on Teaching Evolution." *Journal of Media and Religion* 2, no. 1 (2003): 5–28.

McGreal, Chris. "Sanity or Honour? TV Pundits Invite U.S. Electorate to Choose Sides." *Guardian*, October 26, 2010, 19.

Mencken, H. L. "Mencken Likens Trial to a Religious Orgy, with Defendant a Beelzebub." *Baltimore Evening Sun*, July 11, 1925.

Menuge, Angus. "Who's Afraid of ID?" In *Debating Design: From Darwin to DNA*, edited by William A. Dembski and Michael Ruse, 32–51. Cambridge: Cambridge University Press, 2004.

Miller, Kenneth R. *Only a Theory: Evolution and the Battle for America's Soul*. New York: Penguin, 2008.

Mooney, Chris, and Matthew C. Nisbet. "Undoing Darwin." *Columbia Journalism Review* 44, no. 3 (September/October 2005): 30–39.

MSNBC. "'Hardball with Chris Matthews' for April 21, 2005." Accessed October 5, 2010. http://www.msnbc.msn.com/id/7602221.

National Science Board. "National Science Board Statement on Action of the Kansas Board of Education on Evolution." August 20, 1999. Accessed October 8, 2010. http://www.nsf.gov/nsb/publications/pub_summ .jsp?ods_key=nsb99149.

Neuman, W. Russell, Marion R. Just, and Ann N. Crigler. *Common Knowledge: News and the Construction of Political Meaning*. Chicago: University of Chicago Press, 1992.

Nisbet, Matthew C. "Framing Science: A New Paradigm in Public Engagement." In *Communicating Science: New Agendas in Communication*, edited by LeeAnn Kahlor and Patricia A. Stout, 40–67. New York: Routledge, 2010.

Nisbet, Matthew C., and Chris Mooney. "Framing Science." *Science* 316 (April 6, 2007): 56.

Nord, Warren A. "Intelligent Design Theory, Religion, and the Science Curriculum." In *Darwinism, Design, and Public Education*, edited by John Angus Campbell and Steven C. Meyer, 45–58. East Lansing, MI: Michigan State University Press, 2003.

Numbers, Ronald L. "Darwinism, Creationism and 'Intelligent Design.'" In *Scientists Confront Creationism and Intelligent Design*, edited by Andrew J. Petto and Laurie R. Godfrey, 31–58. New York: W.W. Norton, 2007.

Paley, William. *Existence and Attributes of the Deity, Collected from Appearances of Nature*. London: R. Faulder, 1802.

Pan, Zhongdang, and Gerald M. Kosicki. "Framing Analysis: An Approach to News Discourse." *Political Communication* 10, no. 1 (1993): 55–75.

Pemberton, Tricia. "Professor's Teaching Stirs Campus Debate." *Oklahoma City Oklahoman*, August 8, 2010, 11A.

"Pennsylvania Polka: Santorum Accused of Dancing Around on 'Intelligent Design.'" *Education Week* 26, no. 19 (January 18, 2006): 26.

Peterson, Dan. "What's the Big Deal about Intelligent Design?" *American Spectator* 38, no. 10 (2006): 30–37.

Petto, Andrew J., and Laurie R. Godfrey. "Why Teach Evolution?" In *Scientists Confront Intelligent Design and Evolution*, edited by Andrew J. Petto and Laurie Godfrey, 405–441. New York: W.W. Norton, 2007.

Pigliucci, Massimo. "Evolution, Schmevolution: Jon Stewart and the Culture Wars." In *The Daily Show and Philosophy: Moments of Zen in the Art of Fake News*, edited by Jason Holt, 190–202. Malden, MA: Wiley Blackwell, 2007.

Poole, Steven. *Unspeak*. New York: Grove Press, 2006.

Quammen, David. *The Reluctant Mr. Darwin: An Intimate Portrait of Charles Darwin and the Making of His Theory of Evolution*. New York: W.W. Norton, 2006.

Richards, Jay W. "Why Are We Here? Accident or Purpose?" In *Intelligent Design 101*, edited by H. Wayne House, 131–152. Grand Rapids, MI: Kregel Publications, 2008.

Rosenhouse, Jason, and Glenn Branch. "Media Coverage of 'Intelligent Design.'" *BioScience* 56, no. 3 (March 2006): 247–252.

Ross, Hugh. "Continental Land Mass Growth and the Genesis 1 Chronology." October 8, 2008. Accessed September 6, 2010. http://www.reasons.org/ evolution/theistic-evolution/continental-landmass-growth-and-genesis -1-chronology.

Rothstein, Edward. "Adam and Eve in the Land of Dinosaurs." *New York Times*, May 24, 2007, E1.

Rudoren, Jodi. "Ohio Board Undoes Stand on Evolution." *New York Times*, February 15, 2006, A14.

Ruse, Michael. "The Argument from Design: A Brief History." In *Debating Design: From Darwin to DNA*, edited by William A. Dembski and Michael Ruse, 13–31. Cambridge: Cambridge University Press, 2004.

Ruse, Michael. "Turning Back the Clock." In *Intelligent Design: Science or Religion*, edited by Robert M. Baird and Stuart E. Rosenbaum, 131–151. Amherst, NY: Prometheus, 2007.

Ruse, Michael. *Darwin and Design: Does Evolution Have a Purpose?* Cambridge, MA: Harvard University Press, 2003.

Ruse, Michael. *Charles Darwin*. Malden, MA: Blackwell Publishing, 2008.

Ryan, Michael, and James W. Tankard, Jr. *Writing for Print and Digital Media*. New York: McGraw-Hill 2005.

Scott, Eugenie C. "Creation Science Lite." In *Scientists Confront Intelligent Design and Creationism*, edited by Andrew J. Petto and Laurie R. Godfrey. New York: W.W. Norton, 2007.

Scott, Eugenie C. *Evolution vs. Creationism: An Introduction*. 2nd ed. Westport, CT: Greenwood Press, 2009.

Scott, Eugenie C. "The Once and Future Intelligent Design." In *Not in Our Classrooms: Why Intelligent Design Is Wrong for Our Schools*, edited by Eugenie C. Scott and Glenn Branch, 1–27. Boston: Beacon Press.

Sharpes, Donald K., and Mary M. Peramas. "Accepting Evolution or Discarding Science." *Kappa Delta Pi Record* 158 (Summer 2006).

Sober, Elliott. "The Design Argument." In *Debating Design: From Darwin to DNA*, edited by William A. Dembski and Michael Ruse, 98–129. Cambridge: Cambridge University Press, 2004.

Sotirovic, Mira. "Effects of Media Use on Audience Framing and Support for Welfare." *Mass Communication and Society* 3, nos. 2 and 3 (2000): 269–296.

Stepp, Diane R., and Kristina Torres. "Cobb Gives Up on Evolution Book Stickers." *Atlanta Journal-Constitution*, December 20, 2006, A1.

Strobel, Lee. *The Case for a Creator: A Journalist Investigates Scientific Evidence That Points Toward God.* Grand Rapids, MI: Zondervan, 2004.

Sunlon, Bill. "Professor Says Dover 'Misleads' Students." *Harrisburg (PA) Patriot News*, September 27, 2005, A1.

Tankard, James W., Jr. "The Empirical Approach to the Study of Framing." In *Framing Public Life*, edited by Stephen D. Reese, Oscar H. Gandy, and August E. Grant. Mahwah, NJ: Lawrence Erlbaum Associates, 2001.

Thaxton, Charles, Walter Bradley, and Roger Olsen. *The Mystery of Life's Origin.* Dallas: Lewis and Stanley, 1984.

Torres, Kristina. "Evolution Tags Aimed to Spur Religious Talk." *Atlanta Journal-Constitution*, November 10, 2004, A1.

Trigilio, John Jr., and Kenneth Brighenti. *The Catholicism Answer Book: The 300 Most Frequently Asked Questions.* Naperville, IL: Sourcebooks, 2007.

Tuchman, Gaye. "Objectivity as a Strategic Ritual: An Examination of Newsmen's Notions of Objectivity." *American Journal of Sociology 77*, no. 4 (1972): 660–679.

Tuchman, Gaye. *Making News: A Study in the Construction of Reality.* New York: Free Press, 1978.

Vergano, Dan. "Fossils Offer 'Window' into Human Evolution." *USA Today.* April 9, 2010, A2.

Walekjo, Gina, and Thomas Ksiazek. "Blogging from the Niches: The Sourcing Practices of Science Bloggers." *Journalism Studies* 11 no. 3 (June 2010): 412–437.

Wells, Jonathan. *Icons of Evolution: Science or Myth.* Washington, DC: Regnery Publishing, 2000.

Wexler, Jay D. "From the Classroom to the Courtroom: Intelligent Design and the Constitution." In *Not in Our Classrooms: Why Intelligent Design Is Wrong for Our Schools*, edited by Eugenie C. Scott and Glen Branch, 83–104. Boston: Beacon Press, 2006.

Wilson, James Q. "Faith in Theory: Why Intelligent Design Simply Isn't Science." In *Intelligent Design: Science or Religion*, ed. by Robert M. Baird and Stuart E. Rosenbaum, 49–52. Prometheus: Amherst, NY, 2007.

Witham, Larry A. *Where Darwin Meets the Bible: Creationists and Evolutionists in the Bible.* Oxford: Oxford University Press, 2002.

York, Chance. Decade of Design: Media Framing of "Intelligent Design" as a Religious/Unscientific Concept or a Scientific/Unreligious Concept from 2000 to 2009. Master's thesis, University of Kansas, 2010.

NEWSPAPER SOURCES FOR FRAMING STUDY

"A Battle Prudently Avoided." *Cleveland Plain Dealer*, February 16, 2006, B8.

"Academic Hall of Shame." *Los Angeles Times*, November 20, 2004, B20.

Anderson, Nick "Cecil County Adopts Text Stressing Evolution." *Washington Post*, February 15, 2005, B3.

Badkhen, Anne. "Anti-Evolution Teachings Gain Foothold in U.S. Schools." *San Francisco Chronicle*, November 30, 2004, A1.

Banerjee, Neela. "Christian Conservatives Press Issues in Statehouses." *New York Times*, December 13, 2004, A1.

Barrow, Bill. "Science Law Could Set Tone for Jindal: Academic Freedom, or a 'Trojan Horse'?" *New Orleans Times-Picayune*, June 27, 2008, 1.

Begley, Sharon. "Teaching Evolution at Christian College." *Chicago Sun-Times*, December 13, 2004, 20.

Blumner, Robyn E. "Backward, Christian Soldiers." *St. Petersburg Times*, November 14, 2004, P7.

Carson, Larry, and Larry Williams. "Evolution or Design?" *Baltimore Sun*, December 19, 2004, F1.

Cohen, Laurence D. "The Revolution Won't Be Televised." *Hartford Courant*, December 1, 2004, A15.

"Creationists at the Gate." *Boston Globe*, January 29, 2005, A14.

Dean, Cornelia. "Evolution Takes a Back Seat in U.S. Classes." *New York Times*, February 1, 2005, F1.

"Debate Keeps Evolving." *Columbus Dispatch*, March 14, 2004, 4B.

Dotinga, Randy. "A Who's Who of Players in the Battle of Biology Class." *Christian Science Monitor*, December 7, 2004, 11.

"Fear of Evolution." *Baltimore Sun*, February 9, 2005, 4A.

Fulwood III, Sam. "Ohio Dressing Up Religion as Science." *Cleveland Plain Dealer*, March 13, 2004, B1.

Glanz, James. "Montana Creationism Bid Evolves into Unusual Fight." *New York Times*, February 29, 2004, 1, 22.

"God and Darwin." *Washington Post*, January 24, 2005, A14.

Hirsch, Arthur. "Cecil to Seek Md. Help on Dispute over Evolution." *Baltimore Sun*, February 15, 2005, 6B.

"Intelligent Redesign." *Atlanta Journal-Constitution*, November 10, 2004, 18A.

Jacoby, Susan. "Caught between Church and State." *New York Times*, January 19, 2005, A19.

"Judge Ruled for What's Right." *Denver Post*. June 25, 2006, E4.

Levy, Paul. "Wisconsin District Is Focus of Evolution Debate." *Minneapolis Star-Tribune*, November 10, 2004, B1.

Louis, Errol. "Unintelligent Design: The Loony Right Is Hammering Science Harder Than Ever." *New York Daily News*, December 31, 2004, 43.

"Louisiana's Latest Assault on Darwin." *New York Times*, June 21, 2008, 18.

McClellan, Bill. "Take an Unscientific Approach Toward Missouri's Future." *St. Louis Post-Dispatch*, October 3, 2004, C1.

McNamee, Tom. "Field Sold on Evolution: Theory for Scientists, While Religiously Motivated Critics Have No Faith in It." *Chicago Sun-Times*, June 25, 2007, 24.

Mishra, Raja. "Evolution Foes See Opening to Press Fight in Schools." *Boston Globe*, November 16, 2004, A1.

Parker, Laura. "School Science Debate Has Evolved." *USA Today*, November 29, 2004, A3.

Quick, Susanne. "Theories Other Than Evolution to Be Taught in Grantsburg." *Milwaukee Journal Sentinel*, November 6, 2004, A1.

Riley, John. "A Matter of 'Intelligent Design.'" *Newsday*, January 14, 2005, A10.

Rosen, Laurel. "Board Backs Selection of Biology Text." *Sacramento Bee*, April 25, 2004, N1.

Ruth, Daniel. "Her 'Academic Freedom'? Not Free, Just Dumb." *Tampa Tribune*, April 3, 2008, Metro 1.

Sappenfield, Mark, and Mary Beth McCauley. "God or Science?" *Christian Science Monitor*, November 23, 2004, P11.

Scharrer, Gary. "Evolution Teaching Provision Fails First Test." *San Antonio News-Express*, January 23, 2009, A1.

Stephens, Scott. "Panel OKs Disputed 10th-Grade Biology Plan." *Cleveland Plain Dealer*, March 10, 2004, A1.

Stephens, Scott, and John Mangels. "Ohio Teachers Dance around Darwin vs. Dogma." *Cleveland Plain Dealer*, February 10, 2005, B4.

"Stick with Evolution." *Atlanta Journal-Constitution*, January 14, 2005, A12.

Strauss, Valerie. "Fresh Challenges in the Old Debate over Evolution." *Washington Post*, December 7, 2004, A14.

Stutz, Terrence. "Professors Back Evolution Alone." *Dallas Morning News*, November 18, 2008, 5A.

"Sweating Details, Skewing the Science." *Cleveland Plain Dealer*, November 14, 2004, H2.

Toland, Bill. "Evolutionary Challenge." *Pittsburgh Post-Gazette*, December 15, 2004, A1.

Torres, Kristina. "Court to Weigh In on Evolution Feud." *Atlanta Journal-Constitution*, November 7, 2004, A1.

Torres, Kristina. "Evolution Tags Aimed to Spur Religious Talk." *Atlanta Journal-Constitution*, November 10, 2004, A1.

Van Deerlin, Lionel. "Slowing the Advance of Learning." *San Diego Union-Tribune*, January 5, 2005, B7.

Westneat, Danny. "Institute's Ways Need to Evolve." *Seattle Times*, November 17, 2004, B1.

Index

About the Author

MARK PAXTON is a professor in the Department of Media, Journalism, and Film at Missouri State University, where he has taught since 1995. He has written extensively on First Amendment issues, particularly as they apply to college media and students' freedom of expression, and has been active in the American Civil Liberties Union. Before entering academia, Paxton was a reporter and news editor for the Associated Press and for newspapers in West Virginia and Tennessee.